Surface Science Investigations in Tribology

ACS SYMPOSIUM SERIES **485**

Surface Science Investigations in Tribology

Experimental Approaches

Yip-Wah Chung, EDITOR
Northwestern University

A. M. Homola, EDITOR
IBM Almaden Research Laboratory

G. Bryan Street, EDITOR
IBM Almaden Research Laboratory

Developed from a symposium sponsored
by the Division of Colloid and Surface Chemistry
at the 201st National Meeting
of the American Chemical Society,
Atlanta, Georgia,
April 14–19, 1991

American Chemical Society, Washington, DC 1992

Library of Congress Cataloging-in-Publication Data

Surface science investigations in tribology: experimental approaches / Yip-wah Chung, editor, A. M. Homola, editor, G. Bryan Street, editor.

p.   cm.—(ACS symposium series, ISSN 0097–6156; 485)

"Developed from a symposium sponsored by the Division of Colloid and Surface Chemistry at the 201st National Meeting of the American Chemical Society, Atlanta, Georgia, April 14–19, 1991."

Includes bibliographical references and index.

ISBN 0–8412–2205–3

1. Tribology—Experiments.

I. Chung, Yip-wah, 1950–   . II. Homola, A. M., 1935–   . III. Street, G. Bryan, 1938–   . IV. American Chemical Society. Division of Colloid and Surface Chemistry. V. American Chemical Society. Meeting (201st: 1991: Atlanta, Ga.) VI. Series.

TS1075.S87   1992
621.8′ 9—dc20
              92–1263
              CIP

SD 4/20/92 RL

Foreword

THE ACS SYMPOSIUM SERIES was founded in 1974 to provide a medium for publishing symposia quickly in book form. The format of the Series parallels that of the continuing ADVANCES IN CHEMISTRY SERIES except that, in order to save time, the papers are not typeset, but are reproduced as they are submitted by the authors in camera-ready form. Papers are reviewed under the supervision of the editors with the assistance of the Advisory Board and are selected to maintain the integrity of the symposia. Both reviews and reports of research are acceptable, because symposia may embrace both types of presentation. However, verbatim reproductions of previously published papers are not accepted.

Contents

INDEXES

Preface

IN THE CLASSIC TEXTBOOK *Principles of Tribology* edited by J. Halling, tribology is defined as the science and technology of interacting surfaces in relative motion and of related subjects and practices. This all-encompassing definition reflects accurately the trend of tribology today. Tribology is no longer a subject dealing with bearings and gears. Surface chemists and materials scientists can contribute a great deal to tribology in terms of fundamental understanding of interfacial interactions and development of wear-resistant materials and better lubricants. It is with this spirit that we organized the symposium "Applications of Surface Science and Advances in Tribology." We hope that the symposium will further underscore the importance of long-term basic research in tribology and encourage interdisciplinary interactions among surface chemists, materials scientists, and engineers in this dynamic field.

This book begins with an overview of the problems and opportunities in tribology. The subsequent chapters are organized into two sections. The first deals with the application of surface diagnostic techniques to study the relationship between surface chemistry and tribological properties of organic films, coatings, and solid lubricants. The second section focuses mainly on liquid lubricants and their interactions with surfaces. Because the various surface analytical tools described in this book may not be familiar to all, a survey chapter is included that describes some of the more commonly used surface tools.

The symposium upon which this book was based was sponsored by the ACS Division of Colloid and Surface Chemistry and the National Science Foundation under grant no. MSS–9022666. We would like to thank all the participants and authors of the symposium for their contributions, A. Maureen Rouhi, Cheryl Shanks, and Anne Wilson of the ACS Books Department for overseeing the publication of this book, and J. Larsen-Basse of the National Science Foundation for his support of this project.

YIP-WAH CHUNG
Northwestern University
Evanston, IL 60208–3108

G. BRYAN STREET
IBM Almaden Research Center
San Jose, CA 95120–6099

A. M. HOMOLA
IBM Almaden Research Center
San Jose, CA 95120–6099

November 1991

Overview

Problems and Opportunities in Tribology

What Is Tribology?

Tribology, the study of friction, wear, lubrication, and adhesion, is fast becoming a multidisciplinary science. It has been a traditional discipline in mechanical engineering. However, with the recent rapid advances in the synthesis of new materials, the traditional approach to performance evaluation of each new material and new lubricant introduced into a given mechanical system is no longer adequate. Together with the trend toward tighter tolerances, higher operating efficiencies, loads, speeds, and reliability of mechanical components, we must develop a fundamental understanding of the behavior of materials in sliding and rolling contact, in order to push mechanical systems beyond their present performance envelopes.

Asperity Contacts

In many engineering tribology studies, one is concerned with friction, wear, and lubrication of machine components (e.g., power train components, gears, bearings, etc.). In an effort to minimize wear, it is desirable to operate under lubrication conditions where there is a fluid film between two sliding surfaces, with thickness at least three times that of the surface roughness (1). In this way asperity contacts between two sliding surfaces are minimized, thereby leading to reduction of wear. This is the regime of full-film or hydrodynamic lubrication. However, in general one operates in the mixed-lubrication regime, where part of the load is carried by the lubricant and part by asperities (2). Such asperity interactions can result in the formation of wear particles, which may lead to severe wear of surfaces by ploughing. In addition, unprotected asperities can be the initiation sites for severe adhesive wear (scuffing) and seizure failure.

Lubricant–Surface Interactions

Extreme-pressure (EP) additives such as zinc dithiophosphate (ZDP) are often introduced into a lubricant package to provide protection of asperities. These additives react with the surface to form a chemical film (known as a boundary film) that serves at least two important functions (3): (i) prevention of direct metal–metal contact and hence seizure; (ii) providing a lower shear strength layer to reduce friction. In some operations such as metal-forming (e.g., cold rolling of aluminum), massive plastic deformation occurs at high rates. Whatever boundary films formed prior to deformation are likely to crack under these conditions. New boundary films must be formed rapidly to protect the exposed bare metal surfaces. Therefore, there is a need to investigate what reactions are taking place between the additive and the surface, their kinetics, and the stability of reaction products that provide the friction and wear reduction. The latter issue is important because even under modest sliding conditions (about 1 m/s), frictional heating due to asperity contacts can reach 500 K or higher (4). These boundary layers should be able to withstand such flash temperatures in order to provide protection. Armed with such chemical information and a knowledge of how the asperity temperature depends on contact stresses and sliding speeds, one hopes to determine the limits of operation conditions for which friction and wear are acceptable without performing exhaustive friction and wear experiments (5). This is the ultimate goal of engineering tribology.

Advanced Tribo-Materials and High-Temperature Lubrication

The need to operate at higher engine efficiencies has initiated a lot of interest in the development of new high-temperature light-weight structural materials and high-temperature lubricants. The ideal structural material should have good mechanical strength, chemical inertness, wear resistance, and toughness at the required operating conditions. Many aluminum-based alloys have excellent strength-to-weight ratios, and their wear resistance can be markedly enhanced by introducing ceramic particles into the matrix (6). Development of such metal–matrix composites is a prime area of research in materials science and tribology.

Application of ceramics and ceramic composites as tribo-components is another important research and development activity. These materials satisfy all the requirements of ideal high-temperature materials except toughness. The low toughness of ceramic materials is a major barrier against their widespread use as structural materials. There is a need to develop ceramics with improved toughness (7).

In the area of high-temperature liquid lubricants, there is a lot of interest in perfluoropolyethers (PFPEs) because of their excellent thermal and oxidation stability (8). However, PFPEs decompose rapidly when exposed to clean iron surfaces (9). Proper additives soluble in PFPEs must be developed to prevent direct metal–PFPE interactions. There are other possibilities to handle lubrication at elevated temperatures (e.g., solid lubricants such as molybdenum disulfide, graphite, or the PS200 composite solid lubricant developed at NASA). The major concern in the use of solid lubricants is replenishment. A recent development is vapor-phase lubrication, in which the lubricant is delivered to the interface as a vapor that decomposes to form a sacrificial boundary film (10, 11). An alternative scheme is the use of self-lubricating materials (e.g., carbon-containing materials and titanium dioxide (12, 13).

Wear-Resistant Coatings

Advances in thin-film techniques have major impacts on the synthesis of wear-resistant coatings. Diamond film technology is a good example. However, hardness and wear-resistance are not the only considerations in choosing a given type of coating. The coating must have good toughness and adhesion to the substrate. More important, the coating process must be carried out at a sufficiently low temperature so as not to affect the microstructure and chemical composition of the substrate. As a result, many wear-resistant coatings are deposited by sputtering. In general, tribological properties of these coatings depend on surface preparation and actual deposition conditions. The relationship among tribological properties, processing, and microstructural properties is an important area of tribology.

Applications Beyond Traditional Tribo-Components

Tribological problems are not confined to automotive and aerospace industries. As we enter into the information age, we see the continuing miniaturization of integrated circuits and the availability of hard-disk systems with ever-increasing capacities. In each case, understanding of basic tribology issues is critical to improving the reliability of these systems. In clean rooms used for fabrication of ultra-large-scale integrated circuits, sophisticated air filtration systems can virtually eliminate particulate contamination from external sources; yet, submicrometer particulate contamination is still a major problem. The reason is that machineries operating inside clean rooms can produce submicrometer particles; even the mere act of writing with a regular pencil produces wear particles.

In the operation of computer hard-disk systems, the read–write head flies above the hard-disk surface with a gap spacing of about 0.1–0.15 μ. The head contacts the disk surface during power-up and power-down and at times of mechanical instability. The friction, wear, and interfacial chemistry of these systems in different environments is a subject of intensive ongoing research (14–18). In an effort to increase the storage capacity of hard-disk media, the industry plans to reduce the flying height to 0.05 μ or even to zero (contact recording) while maintaining the same reliability. Therefore, the soft thin-film magnetic media (typically made of Ni–Co alloys) must be protected in some way. A number of solutions are being investigated: (i) reduction of loading on the read–write head from about 100 mN to 0.1 mN; (ii) development of better lubricants; (iii) reduction of surface roughness to minimize asperity interactions; and (iv) development of more wear-resistant protective coatings.

Another emerging area concerned with wear and mechanical failure of tribo-components is medicine. Artificial heart valves and load-bearing implants are subjected to repetitive wear cycles under a corrosive environment; yet they must last a long time. This is one instance where long-term reliability outweighs all other concerns.

Closing Remarks

A few years ago, Rabinowicz at MIT estimated that about 6% of the GNP is lost because of incomplete understanding and application of tribology (19). Therefore, even a modest improvement in friction and wear can lead to substantial economical benefits. Although this survey is by no means complete, the diversity and significance of many tribological problems are clear. With the rapid advances in surface diagnostics, we can routinely study surface composition and lubricant–surface interactions, obtain detailed surface topography, and probe the microscopic details of friction and lubrication. The stage is now set for atomic-scale investigations of many fundamental problems in tribology. Time has come for the surface and materials sciences community to venture into this most fruitful discipline.

References

1. *Principles of Tribology*; J. Halling, Ed.; Macmillan, 1975.
2. D. Dowson, *Boundary Lubrication: An Appraisal of World Literature*; Ling, F. F.; Klaus, E. E.; Fein, R. S., Eds.; Am. Soc. Mech. Eng., 1969; p 229.

3. Rhodes, K. L.; Stair, P. C. *J. Vac. Sci. Technol.* **1988,** *A6,* 971.
4. Ashby, M.; Abulawi, J.; Kong, H. *Trib. Trans.* **1991,** *34,* 577.
5. Shen, M. C.; Cheng, H. S.; Stair, P. C. *J. Tribol.* **1991,** *113,* 182.
6. Fine, M. E.; *Advances in Engineering Tribology*; Chung, Y. W.; Cheng, H. S. Y. W., Eds.; Soc. Trib. and Lubr. Engrs., 1990; p 151.
7. Gangopadhyay, A.; Janon, F.; Fine, M. E.; Cheng, H. S. *Trib. Trans.* **1990,** *33,* 96.
8. Snyder, C. E., Jr.; Gschwender, L. J. *Advances in Engineering Tribology*; Chung, Y. W.; Cheng, H. S., Eds.; Soc. Trib. and Lubr. Engrs., 1990; p 216.
9. Mori, S.; Morales, W. *Trib. Trans.* **1990,** *33,* 325.
10. Lauer, J.; Dwyer, S. *Trib. Trans.* **1990,** *33,* 529.
11. Klaus, E. E.; Jeng, G. S.; Duda, J. L. *Lub. Engrg.* **1989,** *45,* 717.
12. Miyoshi, K.; Buckley, D. H. *Wear* **1986,** *110,* 295.
13. Gardos, M. N.; Hong, H.-S.; Winer, W. O. *Trib. Trans.* **1990,** *33,* 209.
14. See, for example, *Tribology and Mechanics of Magnetic Storage Systems*; Bhushan, B.; Eiss, N. S., Eds.; Soc. Tribol. Lubr. Engrs., vol. 1–7 (1984–1990).
15. Dugger, M. T.; Chung, Y. W.; Bhushan, B.; Rothschild, W. *J. Tribol.* **1990,** *112,* 238.
16. Streator, J. L.; Bhushan, B.; Bogy, D. B. *J. Tribol.* **1991,** *113,* 22.
17. Marchon, B.; Heiman, N.; Khan, M. R. *IEEE Trans. Magn.* **1990,** *23,* 168.
18. Wahl, K. J.; Chung, Y. W.; Bhushan, B.; Rothschild, W. *Ad. Info. Storage Syst.* **1991,** *3,* 83.
19. Rabinowicz, E. *Tribology and Mechanics of Magnetic Storage Systems*; Bhushan, B.; Eiss, N. S.; STLE, 1986; vol. 3, p 1.

YIP-WAH CHUNG
Northwestern University
Evanston, IL 60208–3108

APPLICATIONS OF SURFACE DIAGNOSTIC TECHNIQUES

Chapter 1

Surface Diagnostic Techniques Commonly Used in Tribological Studies

Yip-Wah Chung

Department of Materials Science and Engineering
and
Center for Engineering Tribology, McCormick School of Engineering
and Applied Science, Northwestern University, Evanston, IL 60208

In this paper, we present an overview of several surface diagnostic techniques commonly used in tribological studies. These include Auger electron spectroscopy, X-ray photoemission, high resolution electron energy loss spectroscopy and scanning probe microscopy. For each technique, we discuss its physical basis, experimental aspects and data interpretation.

Many of the studies discussed in this book call for surface analysis before, during and after certain friction and wear experiments. A number of surface diagnostic techniques are available that provide direct information on surface composition, chemical state, molecular identity of adsorbates, surface morphology and structure. In some instances, instruments for friction and wear experiments are integrated with those for surface analysis so that both types of measurements can be done in situ. This approach, while most desirable, is not always required. Post-testing surface analysis is sufficient in many cases.

Some of the techniques described here require placing the test specimen in an ultrahigh vacuum chamber. Since it is generally undesirable to place "dirty" specimens in a UHV chamber, there is a limit to what type and size of specimens one can examine by these techniques. However, as we see in this book, practical specimens of different types have been successfully investigated by these techniques.

In the following section, we will discuss four commonly used techniques, viz. Auger electron spectroscopy, X-ray photoelectron spectroscopy, high-resolution electron energy loss spectroscopy and scanning probe microscopy.

0097–6156/92/0485–0002$06.00/0

2. Auger Electron Spectroscopy

2.1 Auger Electron Emission

Consider atoms in a solid being bombarded by electrons or X-rays. When the energy of the incident radiation is larger than the binding energy of electrons in some core level, say the K-shell, there is a certain probability that the K-shell electron is knocked out of the atom. The resulting ion is in an excited state. An electron from a higher energy shell, say the L shell, falls down to fill the K-vacancy. The excess energy $E_K - E_L$ can then be released as X-rays or given to a third electron, say in the M shell of the same atom (E_i is the binding energy of electrons in the ith shell with respect to the Fermi level). The former process gives rise to X-rays of energy equal to $E_K - E_L$ and is the basis of electron beam microprobe. The latter process (fig.1) results in the emission of an electron with an energy equal to $E_K - E_L - E_M - W$, where W is the work function of the surface and is known as a KLM Auger transition. The energy of the emitted electron (Auger electron) is characteristic of the parent atom. Therefore, measurement of Auger electron energies constitutes a method of element identification.

In order for this technique to be useful in surface analysis, we would like to excite Auger electron emission with kinetic energies in the range of 50 to 1500 eV. In this energy, the electron mean free paths in typical solids is 4-15 A. Such small mean free paths are the basis of the surface sensitivity for this technique.

2.2 Experimental Aspects

(a) excitation source. Auger electron transitions can be excited by electrons or X-rays. We will postpone the discussion of X-ray excitation to the next section. Electrons are usually generated by thermionic emission from filament materials such as tungsten or lanthanum hexaboride (LaB_6). Lanthanum hexaboride is a stable low work function material which allows one to extract a larger electron beam current than tungsten at the same temperature and beam size. Such high brightness performance is required in high resolution Auger studies. Some newer Auger systems are based on field emission electron guns. Electron guns are designed to have electron spot sizes ranging from a millimeter to less than 500 A and are operated at accelerating voltages 2-30 kV.

(b) electron analyzers. In modern surface analysis, a band-pass analyzer such as a cylindrical mirror analyzer or concentric hemispherical analyzer (1,2) is used. Since Auger signals are normally small peaks on top of a large but smooth secondary background, the Auger spectrum is usually recorded as the first derivative of the electron energy distribution N(E) to enhance spectral features.

2.3 Sensitivity of Auger Electron Spectroscopy

(a) element. Auger analysis of solid surfaces can detect all elements except hydrogen and helium.

(b) number. The actual number sensitivity, i.e. the lowest concentration detectable, depends on several experimental parameters, viz. accelerating voltage, electron beam current, Auger cross section of the peak one is looking at, the electron analyzer transmission, and the data acquisition time or integration time. Under typical conditions, the lowest detectable limit ranges from a few percent to 0.1 percent of a monolayer.

(c) surface. The surface sensitivity of AES depends on the electron kinetic energy to be measured. The probability for an electron to travel a distance t in the solid without any inelastic collision is exp(-t/L), where L is the mean free path. Therefore, 95 % of the Auger signal comes from within 3L of the surface, assuming that Auger electrons are collected at normal exit from the surface. This sampling depth of 3L can be reduced to 3L cosΘ by detecting Auger electrons at an angle Θ from the sample surface normal. Obviously, this enhancement does not work too well with rough surfaces.

2.4 Chemical Effects

When an atom is placed under different chemical environments, two possible changes may result in the Auger spectrum :

(a) shift of the Auger energy peak position. This is caused by shifts of electron energy levels due to electron transfer to or away from the atom of interest.

(b) change of the Auger peak shape. This can be caused by the change in the valence band structure. Also, as the Auger electron exits from the solid, it may undergo energy loss collisions resulting in satellite features. Change in chemical environments can result in change in these features and thus the overall shape of the Auger peak, which can be useful in studying chemical state of atoms on surfaces.

It is possible to obtain chemical state information from these peak position and peak shape changes. However, since three electrons are involved in the Auger emission process, the interpretation is less than straightforward.

2.5 Profile Analysis

Sometimes, one is interested in the depth distribution of elements. This can be determined by removing the surface atoms by inert gas (usually argon) ion sputtering and simultaneously analyzing the surface composition using Auger electron spectroscopy. However, the sputtering rates of different components of the solid can be different so that concentration profile obtained by sequential sputtering and Auger analysis has to be interpreted carefully.

2.6 Scanning Auger Microprobe

In many tribological studies, one requires some reasonable spatial resolution to analyze wear tracks and other microscopic features. This can readily be accomplished by using a fine probing electron beam for exciting Auger transitions. Commercial scanning Auger microprobe instruments are now available with electron probe sizes less than 500 A.

There are two major precautions in using the scanning Auger microprobe. One is the potential for electron damage. In the spot mode (i.e. the beam rests at one spot), the current density can be high enough to cause surface damage. Another precaution is interpretation of absolute signal intensities. When analyzing a rough surface, Auger electron intensity can be enhanced due to edge effects or suppressed due to shadowing. One correction scheme involves normalizing the Auger intensity by the secondary electron background.

2.7 Quantitative Analysis

There are several reasons for Auger electron spectroscopy to become such a popular surface analytical tool. It is relatively easy to use. Interpretation is usually straightforward. Element identification is easy. Auger derivative spectra are compiled for almost all elements in the periodic table (3). It is just a matter of matching peaks in order to have a complete identification of elements on the surface.

Quantitative analysis is more difficult. The first order approach to quantitative analysis is the use of relative sensitivity factors (3). The relative sensitivity factors are obtained by measuring the relative intensities of selected Auger peaks of different elements under identical conditions. The reason that this is only a first order approach is that electron mean free path, Auger ionization probability and backscattering of electrons are dependent on the matrix. For further details, see ref.(4).

3. X-ray Photoelectron Spectroscopy

3.1 One-Electron Description of the Photoelectric Effect

Take a solid whose energy levels are shown in Fig.2. The tightest bound electrons reside in core levels (K and L in this case). The outermost electrons of the solid form a band with a certain occupied density of states distribution up to the Fermi level E_F. On illumination of the solid with photons of energy greater than the work function of the solid ($E_{vac} - E_F$), electrons are excited from these levels and can be ejected from the surface. The energy distribution of these photoelectrons has a one-to-one correspondence with the initial state energy distribution. Note that because of strong electron-electron interaction in a solid, some photoelectrons interact with electrons in the solid to produce secondary electrons on their way to the solid surface. These secondary electrons are superimposed on the photoelectron spectrum as a smooth continuous background.

Consider the case of a core electron of binding energy E_B with respect to the Fermi level E_F (i.e. it is located at an energy E_B below the Fermi level). When this core electron is ejected by a photon of energy $h\nu$, the kinetic energy E_{Kin} of the resulting photoelectron is given by

$$E_{Kin} = h\nu - E_B - W$$

where E_{Kin} is referenced to the vacuum level of the specimen E_{vac}, and W is the work function of the specimen. This equation describes the photoemission

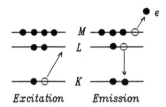

Fig.1 Mechanism of Auger electron emission

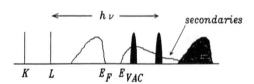

Fig.2 Illustration of the photoemission process. E_F - Fermi level; E_{VAC} - vacuum level.

process in the simplest approximation (the 'frozen orbital' approximation) and does not take into account the fact that upon photoemission, the species left behind has a $+e$ charge presenting a different potential to the outgoing photoelectron and the surrounding electrons.

3.2 Experimental Aspects

The most commonly used X-ray sources are Al (1486.6 eV) and Mg K α (1253.6 eV) lines. These X-ray lines are narrow (less than 0.9 eV) and thus provide excellent energy resolution. As in Auger electron spectroscopy, photoelectrons are detected by a band-pass energy analyzer. Unlike Auger electron spectroscopy, it is non-trivial to perform high spatial resolution or small-spot XPS. One approach is to use a crystal or mirror to focus the X-ray beam into a small spot. Another approach, which is in commerical use, is to use electron optics at the entrance of the electron energy analyzer to collect photoelectrons from a small area on the specimen surface. Resolution down to 10 microns has been demonstrated.

3.3 Element Identification

The X-ray photon energy from the Mg or Al K α line is sufficient to excite electron emission from core levels. Subsequent relaxation within the excited ion can result in the emission of Auger electrons as described in section 2. There-fore, an X-ray photoelectron spectrum contains Auger information for element identification purposes. In addition, core levels of atoms have well- defined bind-ing energies. Therefore, element identification can also be accomplished by locating the energy position of these core levels. Because of the ability of X-ray photoelectron spectroscopy to identify elements, this technique is also known as ESCA (electron spectroscopy for chemical analysis).

The typical number sensitivity of ESCA is about the same order of electron-excited Auger electron spectroscopy, i.e. $\sim$ 0.1-1% of a monolayer. The ad-vantage of XPS is two-fold: (i) X-ray photons appear to be less damaging to sur-faces and adsorbates than electrons; (ii) An X-ray photoelectron spectrum con-tains chemical state information that is sometimes more easily interpreted than the corresponding Auger electron spectrum.

3.4 Chemical Shift

Consider a free atom of sodium whose electronic configuration is $1s^2 2s^2 2p^6 3s^1$. Let us assume that sodium participates in a certain chemical reac-tion in which the outermost valence electron is removed, e.g., Na reacting with Cl to give $Na^+ Cl^-$. All the remaining electrons in the sodium ion will be moving in a different potential. In particular, the electrostatic potential seen by say the 1s electrons will increase because of the net positive charge of the whole system. The binding energy of the 1s level will increase. This is known as chemical shift. For transition metals which exhibit multiple oxidation states, a monotonic cor-relation between binding energy shift and the oxidation number exists. For ex-ample, the binding energies of the $2p_{3/2}$ core level of Cu are 932.8, 934.7 and 936.2

eV in pure Cu, Cu_2O and CuO respectively. Chemical shifts are typically on the order of eV's.

3.5 Relaxation Shift

A phenomenon called relaxation complicates the simple chemical shift analysis. For example, when an inert gas atom is implanted into different metals such as gold, silver and copper, the measured binding energy of a given core level of the inert gas atom depends on the surrounding medium in spite of the absence of chemical bonding between the inert gas and the metal (5). In the photoionization process, the outgoing photoelectron and the electron vacancy or hole left behind have an attractive interaction. Electrons in the surrounding medium relax towards this hole, thus screening the photoelectron- hole attractive interaction. This relaxation results in a higher measured kinetic energy of the photoelectron (or smaller apparent binding energy) than when relaxation is absent. Since relaxation depends on the medium, the measured binding energy also depends on the medium. Such a shift in the absence of chemical bonding is known as relaxation shift.

Because of these relaxation effects, the same chemical species adsorbed on different surfaces exhibiting the same chemical state (e.g. phosphorus adsorbed on Al and Fe) may have different core level positions as observed by XPS. Unfortunately, assignment of chemical states based on core level positions without proper reference is still a common practice.

4. High Resolution Electron Energy Loss Spectroscopy

4.1 Introduction

Consider a low energy electron approaching a given surface. The charge on the approaching electron induces an electric field on the surface. The electric field intensity changes with time as the electron impacts and scatters off the surface. This time-dependent electric field can excite certain atomic vibrations on the surface. The quantum of this surface vibration $\hbar\omega$, by conservation of energy, must be taken from the incident electron beam. Therefore, the energy distribution of scattered electrons would show a peak at an energy $\hbar\omega$ below the elastic peak (zero-loss peak), i.e. energy positions of these loss peaks give directly the vibrational energies. Fig.3 is a vibrational loss spectrum of a titanium foil oxidized under 3×10^{-6} Torr oxygen at 673 K for 5 minutes. The most prominant peak at 790 cm^{-1} is due to optical phonons from TiO_2 (6). Peaks at 1580 and 2380 cm^{-1} are due to multiple losses. Note that 1 meV energy loss is equivalent to 8.065 cm^{-1}.

4.2 Experimental Aspects

Most surface vibration measurements cover the energy range from 50 meV to 400 meV (400 cm^{-1} to 3200 cm^{-1}, or 125 microns - 3 microns). High resolution EELS can be considered to be the electron analog of Raman scattering. The key instrumentation requirements in high-resolution EELS are: (i) the incident electron beam must be monochromatized to give a line width of 10 meV ($\sim$ 80 cm^{-1}) or less; and (ii) the electron analyzer and electronics must provide sufficient energy resolution so that peaks can be located with an accuracy of 1-2 meV.

HREELS is sensitive to the presence of a few percent of a monolayer of most adsorbates on surfaces. For certain adsorbates (e.g. CO), the sensitivity can be one to two orders of magnitude better. Moreover, it can detect H (from the H-substrate stretching vibration) while other techniques such as Auger electron spectroscopy cannot. Normally, an incident electron beam with energies 1 - 10 eV and currents $\sim$ 1 nA is used so that this spectroscopy technique provides a non-destructive tool for studying atomic and molecular adsorption.

4.3 Excitation Mechanisms

The excitation of surface vibrations occurs through two mechanisms. The first mechanism is long-range dipole scattering. As the incident electron approaches the surface, an image charge (positive) is induced on the surface. The incident electron and its image act together to produce an electric field perpendicular to the surface. Therefore, only vibrations having dynamic dipole moment components normal to the surface can be excited by this mechanism. This is sometimes known as the dipole selection rule in HREELS. Theory shows that the vibrational loss intensity due to dipole scattering peaks in the specular direction (i.e. angle of incidence = angle of scattering). The second mechanism is short-range impact scattering. An incoming electron can interact with adsorbates in a short-range manner near each atom or molecule to excite vibrations. In this case, the scattering cross section depends on the microscopic (atomic) potential of each scatterer. In contrast to dipole scattering, electrons scattered this way are distributed in a wide angular range, and the surface dipole selection rule does not apply.

5. Scanning Probe Microscopy

5.1 Introduction

Scanning probe microscopy refers to a class of surface diagnostic techniques that operate by scanning a fine probe on a specimen surface (without touching the surface). The first such instrument is the topographiner developed by Young in the early 70's (7), followed by the now well-known scanning tunneling microscope. The scanning tunneling microscope (STM) was invented by Rohrer and Binnig of IBM's Zurich Research Laboratory in Switzerland in 1982 (8). One of the most interesting aspects of this new microscopy technique is its ability to perform high resolution imaging of surfaces over a relatively large scanning range both in the horizontal and vertical direction. What is more significant is that such high resolution is achieved in vacuum, air and liquid environments, thus making this new technique extremely convenient to use for practical specimens.

5.2 Principle of STM Imaging

Consider a sharp conducting tip brought to within one nm of a specimen surface (fig.4). Typically, a bias of 0.01 - 1 volt is applied between the tip and the specimen. Under these conditions, the tip-surface spacing s is sufficiently small that electrons can tunnel from the tip to the specimen. As a result, a current I flows across this gap which can be shown to vary with s as follows :

$$I \sim \exp(-10.25\sqrt{\emptyset}\, s)$$

where $\emptyset$ is the effective work function in eV (3-4 eV for most systems) and s is in nm. One can see that if the tip-surface spacing is increased (decreased) by 0.1 nm, the tunneling current will decrease (increase) by about a factor ten for an effective work function of 4 eV.

One can then exploit this sensitive dependence of the tunneling current I on the tip-surface spacing for topographic imaging as follows. In scanning the tip horizontally across the specimen, any change in the tip-surface spacing results in a large change in the tunneling current I. One can use some feedback mechanism to move the tip up or down to maintain a constant tunneling current. According to the above equation, this implies that one is maintaining a constant tip-surface spacing (assuming constant $\emptyset$). In other words, the up-and-down motion of the tip traces out the topography of the surface, analogous to the conventional technique of stylus profilometry, except that the tip never touches the surface in STM. This is known as constant current imaging, the most common imaging mode used in scanning tunneling microscopy.

Because of the proximity of the tip to the surface and the nature of tunneling, the tunneling electron beam diameter can be very small. For a tunneling junction with a typical work function of 4 eV, the full-width at half-maximum is approximately equal to $0.5\sqrt{z}$ (9), where z is the sum of R, the local radius of curvature of the tip and s, the tip-surface spacing in nm. For example, for R = 0.2 nm, and s = 0.5 nm, the electron beam diameter is on the order of 0.4 nm. This implies that the tunneling current is self-focussed into a region with atomic dimensions. Scanning tunneling microscopy has been demonstrated to yield atomic resolution in many cases.

5.3 Experimental Aspects

(a) Coarse motion control

In order to bring the tip to within tunneling range, one must move the tip over macroscopic distances (hundreds of microns) with a precision of several tens of nm. Several schemes are possible. In the original work by Binnig and Rohrer, they used a piezoelectric inchworm (more on piezoelectric materials later). Some designs are based on purely mechanical means, all using fine-thread screws. One example is shown in fig.5. Consider a conventional 80-pitch screw, i.e. the screw advances by one inch (2.54 cm) after 80 turns. This translates into a motion advance of about 880 nm for one degree of screw rotation. Using a cantilever beam with a velocity ratio of 10, one achieves a precision of 88 nm for one degree of screw rotation.

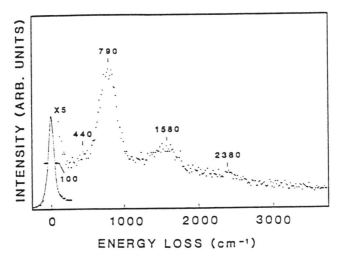

Fig.3 High resolution electron energy loss spectrum of a partially oxidized titanium foil, showing the optical phonon vibration at 790 cm^{-1} and its harmonics.

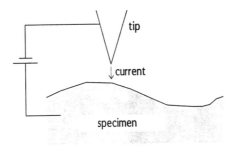

Fig.4 Flow of current across a vacuum gap due to electron tunneling between tip and specimen surface.

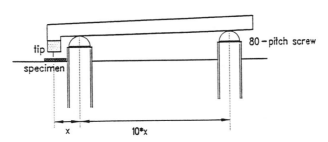

Fig.5 Example of coarse approach using a fine thread screw and the reduced velocity ratio of a cantilever beam.

(b) Fine motion control

During tunneling and feedback control, the precision required for tip positioning relative to the specimen surface must be better than 0.1 nm. This is achieved by piezoelectric positioners. Piezoelectric materials expand or contract upon the application of an electric field. Lead zirconium titanate (PZT) is the material of choice in the STM community. Most STM's are designed with a scanning range from one to ten microns, with some commercial versions going beyond 100 microns. Since voltages can be controlled and monitored in the submillivolt level easily, subnanometer control can be readily attained.

Two types of STM scanners are being used. In one design, three-axis scanning is accomplished using three separate pieces of piezoelectric bars held together in an orthogonal arrangement. In another design, a single piezoelectric tube with four separately biased quadrants provides the capability of three-axis scanning (10). The major advantage of the tube scanner is its improved rigidity.

(c) Tip preparation

Two types of tip materials are widely used, viz. tungsten and platinum alloys (e.g. Pt-Ir and Pt-Rh). Tungsten is strong and can be fabricated into a sharp tip easily, but it tends to oxidize rapidly in air. On the other hand, Pt alloys are stable in air, but they may not survive occasional tip crashes on surfaces.

Several methods can be used to create sharp tips of these materials. These include electro-polishing, cutting and grinding, momentary application of a high bias voltage (a few volts), or simply waiting for a few minutes after setting up in the tunneling configuration. In order to image rough surfaces, tips with large aspect ratios should be used. For further details, see ref. (11).

(d) Vibration isolation

Most STM's are supported using damped springs, air tables or stacked stainless steel plates separated by viton dampers. The goal in all these designs is to keep the tip-surface spacing immune to external vibration. The general design is that one should support the STM on a soft platform and design a microscope with high rigidity.

(e) Data acquisition and analysis

In a typical experiment, a bias of 0.01 - 1 volt is applied between the tip and the specimen. The tunneling current so obtained is then compared with a preset value (typically 1 - 10 nA). The error signal then drives a feedback circuit whose output is used to control a fast high voltage operational amplifier which feeds voltage to the Z electrode of the scanner (which moves the tip perpendicular to the surface). At the same time, raster-scanning is accomplished by using two digital-to-analog converters to control the output of two high voltage operational amplifiers feeding voltages to the X and Y electrodes of the scanner. At each step, the Z voltage required to maintain a constant tunneling current is read by the computer via an analog-to-digital converter (either through AC coupling or potential dividers). This Z voltage, as discussed earlier, corresponds to the surface height at that XY location. This information can then be displayed in real time as gray level images on an analog monitor or stored as two dimensional integer arrays which can later be retrieved for further processing.

5.4 Limitations of STM and Solutions

There are two major limitations of the STM. First, the specimen surface must be reasonably conducting. Under typical operation conditions, the resistance of the gap separating the tip and the specimen is on the order of 10 megaohms (e.g. tunneling at 1 nA under a bias of 10 mV). "Reasonably conducting" means that the resistance of the electrical path from the specimen to the return circuit should be small compared with 10 megaohms. This rules out many ceramic and polymer materials from consideration. One simple solution is to put a conduction coating (e.g. gold) on such surfaces, assuming that the coating faithfully reproduces the surface topography of the substrate. Second, the STM suffers from limited scanning range. The maximum scan range in use today is about 100 microns.

5.5 Atomic Force Microscopy

In scanning tunneling microscopy, the tip-surface spacing is sensed by the tunneling current. This sensing technique does not work too well for highly resistive materials. An alternative scheme to sense the tip-surface spacing is by measuring the force of interaction between the tip and the specimen surface. This is the basis of atomic force microscopy. The tip is normally part of a small wire or microfabricated cantilever. The force of interaction between the tip and the surface results in a deflection of the cantilever. In most modern designs, the cantilever deflection is sensed either by detecting the reflection of a light beam from the back of the cantilever or by optical interferometry. Under appropriate operation conditions, atomic resolution can be achieved. One important strength of the AFM is its ability to obtain images from insulator surfaces. In addition, one can readily adapt an atomic force microscope either as a micro-tribometer (studying friction) or as a nanoindentor (studying surface mechanical properties). For further details, see ref.(12).

6. Some Acronyms

AES	Auger electron spectroscopy
AFM	atomic force microscopy
ESCA	electron spectroscopy for chemical analysis
HREELS	high resolution electron energy loss spectroscopy
SAM	scanning Auger microprobe
STM	scanning tunneling microscopy
XPS	X-ray photoelectron spectroscopy

REFERENCES

1. G. Ertl and J. Kuppers, Low Energy Electrons and Surface Chemistry, Verlag Chemie (1974).
2. H. Ibach, Electron Spectroscopy for Surface Analysis, Springer-Verlag (1977).
3. L. E. Davis, N. C. MacDonald, P. W. Palmberg, G. E. Riach and R. E. Weber, Handbook of Auger Electron Spectroscopy, Physical Electronics Industries (1976).
4. T. E. Gallon, Electron and Ion Spectroscopy of Solids, Plenum Press, edited by L. Fiermans, J. Vennik and W. Dekeyser, p. 230 (1978).
5. P. H. Citrin and D. R. Hamann, Chem. Phys. Lett. 22, 301 (1973).
6. J. A. Gates, L. L. Kesmodel and Y. W. Chung, Physical Review B23, 489 (1981).
7. R. Young, J. Ward and F. Scire, Rev. Sci. Instrum. 43, 999 (1972).
8. G. Binnig, H. Rohrer, Ch. Gerber and E. Weibel, Phys. Rev. Lett. 49, 57 (1982).
9. J. Tersoff and D. R. Hamann, Phys. Rev. B31, 2 (1985).
10. G. Binnig and D. P. E. Smith, Rev. Sci. Instrum. 57, 1688 (1986).
11. A. J. Melmed, J. Vac. Sci. Technol. B9, no. 2, Part II, p.601 (1991).
12. D. Rugar and P. Hansma, Physics Today, vol. 43, no. 10, p.23 (1990).

RECEIVED November 26, 1991

Chapter 2

Chemical and Tribological Behavior of Thin Organic Lubricant Films on Surfaces

Benjamin M. DeKoven[1] and Gary E. Mitchell[2]

[1]Central Research and Development, Materials Science and Development Laboratory, Dow Chemical Company, 1702 Building, Midland, MI 48674
[2]Michigan Division Research and Development, Dow Chemical Company, 1897 Building, Midland, MI 48667

Novel studies involving surface chemistry and friction behavior in ultrahigh vacuum (UHV) of thin films of two novel cyclophosphazene lubricants on clean and oxidized M50 stainless steel surfaces are reported. Results using HREELS, XPS, and *in situ* friction to understand the surface chemical properties are presented and compared with similar experiments conducted using polyphenyl ether. Evidence for tribochemistry of hexa (3-trifluoromethylphenoxy) cylcotriphosphazene similar to that observed previously for polyphenyl ether was found. In contrast no evidence for tribochemistry was seen for hexa (3-methylphenoxy) cylcotriphosphazene under similar conditions. The only difference between these two molecules is fluorination of the methyl group on the phenoxy ring. Thermally induced chemistry was demonstrated on the 'clean' and oxide covered surfaces for both phosphazenes after heating to > 470 K. The tribochemistry is not believed to be thermally induced due to the low loads and sliding speeds. The possibility that the reactions are due to mechanochemistry (mechanical bond breaking) is discussed. The (3-trifluoromethylphenoxy) cylcotriphosphazene fluid reacted both at the aromatic rings, as did polyphenyl ether, and at the CF_3 group. Surprisingly, (3-methylphenoxy) cylcotriphosphazene showed no reactivity of aromatic rings, but thermally induced cleavage of the aryl-O bond was found to occur. Chemisorption of (3-methylphenoxy) cylcotriphosphazene also led to changes in the HREELS spectra which are interpreted in terms of preferred orientation of the phosphorous-nitrogen ring.

The interfacial properties of lubricant molecules are important in many facets of lubrication especially in the boundary layer and extreme pressure regimes. Understanding the chemical and tribochemical interactions of potential lubricant molecules with solid surfaces should speed the development of the next generation of high performance lubricants. Despite the expected usefulness of this information, the systematic study of the interaction of these fluids with surfaces from either a chemical or a mechanical perspective has just begun. The lubricant / surface interface is especially important in the boundary lubrication regime (1-3). Recent work has aimed at understanding the interactions of perfluoropolyalkyl ethers (PFPAE) with surfaces using both model fluorinated ethers (4-8) and the actual fluids (9). Recently, DeKoven and Mitchell reported results of studies of thin films of polyphenyl ether (PPE) with clean and oxidized M50 steel surfaces (10). Aside from our work and the recent work reported by Herrara-Fierro et al. involving reactions of PFPAE on clean steel and Al (9), there is no public information on surface studies involving actual lubricants.

In this work the emphasis was on understanding the interfacial aspects of the tribochemical properties of high temperature lubricant molecules. Two interesting questions which can be addressed by surface science studies are: (1) What is the nature of the chemical interactions between thin layers of adsorbed lubricants with both clean and oxidized surfaces? and (2) Do enhanced surface chemical reactions occur at liquid/metal interfaces under the influence of friction? If so, are the reaction products due to flash thermal excursions or other results of friction such as mechanically induced bond breaking (mechanochemistry)? One result of friction is to cause the protective oxide layer on the metal surfaces to be scraped away and expose an active metal surface. In order to separate tribochemical effects into thermochemical and mechanochemical components, surface chemical and friction studies were conducted on surfaces prepared nearly atomically clean in UHV. It is believed that understanding the catalytic surface degradation of organic liquids will provide insight into the nature of tribochemical interactions.

Here we present new results emphasizing a unique approach towards understanding these interactions by investigating the lubricant / solid interface in ultrahigh vacuum (UHV). The thermal surface chemistry was investigated using high resolution electron energy loss spectroscopy (HREELS) and X-ray photoelectron spectroscopy (XPS). The tribochemical interactions were probed using XPS following in situ (in UHV) friction measurements (2,11-13). In this paper the surface chemical and tribochemical properties of three fluids with M50 steel surfaces are examined. Two novel aryloxycyclotriphosphazenes were investigated and the results were compared with similar studies using a polyphenyl ether, described in detail elsewhere (10). The molecular structures of PPE and both hexa (3-trifluoromethylphenoxy) cylcotriphosphazene (HTFMPCP) and (3-methylphenoxy) cylcotriphosphazene (HMPCP) are shown in Figure 1. The difference between these two phosphazene molecules is the fluorination of the methyl group on the phenoxy ring. The synthesis and physical property characterization of the substituted aryloxycyclotriphosphazenes has recently been described by Nader et al. (14). Basic physical properties of PPE have been reported as well (15-17). All three fluids are strong candidates for high temperature lubrication applications.

Experimental

Surface Analysis. Two separate UHV chambers were used. The first chamber was a VG ESCA Lab II XPS / AES spectrometer equipped with a custom preparation chamber and a rapid specimen entry lock. Small area XPS (viewing area diameter of 1×10^{-3} m) was used to investigate M50 steel surfaces before and after in situ friction measurements. Surfaces were cleaned in UHV at room temperature by ion bombardment (model AG60, VG Inc., Danvers, MA) at 5 keV Ar+ with beam current of ~3 µA/cm². No annealing was performed after sputter cleaning. Specimen heating used a VG heatable stub. The specimens were attached to the heatable stubs using ceramic cement (Aremco Co., Ossining, NY) that was sometimes mixed with graphite to reduce sample charging during the XPS measurement. In cases where the graphite was not used, sample charging occured during XPS measurements. The magnitude of the

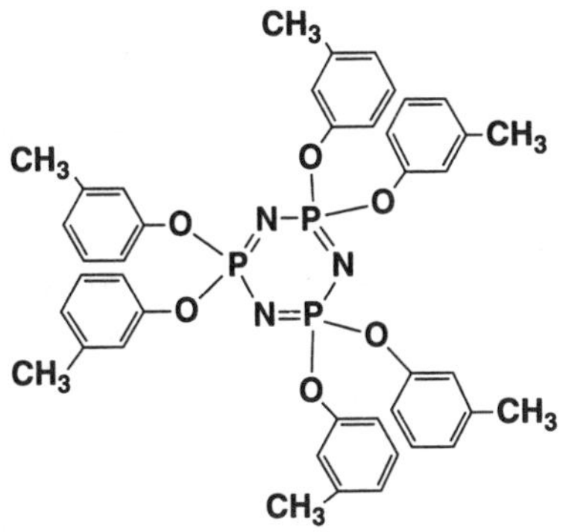

HMPCP

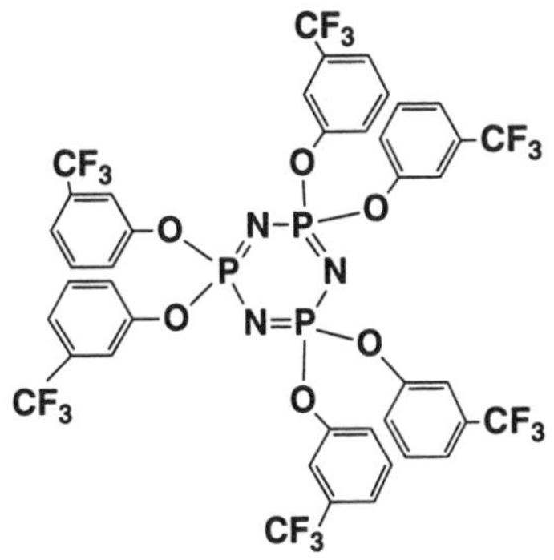

HTFMPCP

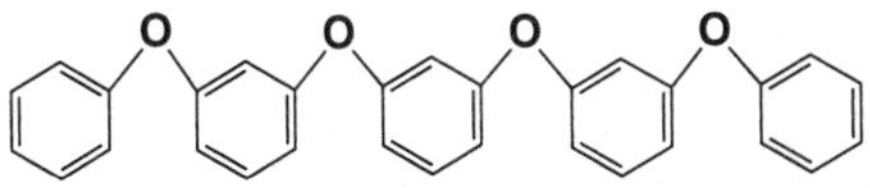

PPE

Figure 1. A pictorial representation of the structures of the three model lubricant molecules (HMPCP, HTFMPCP, and PPE) used in this study.

charging was determined by assigning the large, lowest binding energy C (1s) feature to hydrocarbon at 285.0 eV. The temperature was measured indirectly from the heating current. The relationship between current and temperature was established using a bare stub on which a thermocouple could be mounted directly. The error in the temperature determined from the heating current is estimated to be ± 40 K.

The second UHV system was equipped with high resolution electron energy loss spectroscopy (HREELS, LK Technologies, Bloomington, IN). The HREELS spectra were obtained using 3.5 eV excitation and typically 10^{-10} A (1×10^{-4} A/m^2) beam current. The spectral resolution as determined from the FWHM of the elastic peak ranged from 33-45 cm^{-1}. All spectra reported here were measured in the specular direction using a total scattering angle of 140 °.

Lubricant sources, Dosing Lubricants in UHV, and Film Thickness. The fluids used in this study all had very low vapor pressures (< 10^{-10} Pa) at room temperature (*18*). They were conveniently dosed onto cleaned surfaces in vacuum using a specially prepared evaporation source. The source consisted of tantalum foil disk (75 μm thick and 0.01 m diameter) spot welded onto a 500 μm diameter tantalum wire. A type K thermocouple was spot welded to the back of the disk. The disk was heated resistively and temperature was controlled to within ±1.0 K with an RHK Model TM310 temperature programmer. This was important for dosing reproducibility since the vapor pressure changes by about a factor of 0.3 /K in the 373 K range. For some fluids, in particular the PPE which has high surface free energy (*10*), wetting of the foil was found to be difficult. Nickel mesh (Buckbee Mears Co., Minneapolis, MN) was spot welded to the front face of the source so that capillary action would cause more uniform wetting of the source and consequently more uniform dosing across the surface of samples. With the evaporator placed 0.01-0.02 m from the sample, convenient dosing rates of ~ one monolayer per minute were achieved with temperatures of 350-400 K.

For experiments performed in the XPS chamber, the thickness of the films were measured by the attenuation of the Fe (2p) intensity. The calibration of the XPS measurement was achieved by performing ellipsometry in air (models L118 and L126B, Gaertner Scientific Co., Chicago, IL) on selected surfaces and is described below. The lubricant film thickness was obtained by XPS using the relationship (*19*)

$$I = I_0 \exp[-D_A / \lambda_A], \qquad (1)$$

where I is the XPS intensity of fluid covered substrate, I_0 is the intensity of the clean substrate, D_A is the lubricant film thickness, and λ_A is the inelastic electron mean free path in the fluid for a given kinetic energy electron. The mean free path was determined using D_A measured with ellipsometry and I/I_0 from XPS. The following procedure has been used to determine λ_A which then allows for subsequent determination of D_A by making an XPS intensity measurement. First the lubricant is evaporated onto an *in situ* oxidized M50 steel surface. The XPS Fe (2p) intensity was measured before and after evaporation. Ellipsometry was then used to obtain the optical constants of the lubricant / steel interface in air (*20*). The ellipsometry data was converted to D_A via the method and computer program of McCrackin (*21*). The constants used for the uncovered oxide surface were those measured for UV / ozone cleaned surfaces in air. Although the surfaces oxidized in UHV have about 10 times less carbon on the surface prior to dosing than UV / ozone cleaned surfaces, this amount of C is still small and should not have a large effect on the optical constants (*20*). This is, however, an unavoidable circumstance since ellipsometry in UHV was not available for these film thickness determinations. Measurements were conducted on at least 4 different film covered surfaces for each lubricant ranging in thickness from 40 - 120 Å. The following mean free paths were determined using either HMPCP or HTFMPCP films on M50 steel surfaces: λ_C (1200 eV) = 50 ± 7 Å, λ_O (950 eV) = 31 ± 7 Å, and λ_{Fe} (750 eV) = 30 ± 10 Å. The mean free paths were the same, within experimental error amongst the fluids and agree with previously reported values on other organic materials (*22*). The wetting of the cyclotriphosphazene fluids on M50 steel was not specifically investigated in these studies. Previously the wetting of PPE on M50 was found to be variable and dependent upon surface pretreatments such as UV / ozone cleaning (*10*). Because of the good agreements in the electron mean free paths between fluids (*10, 22*) we do not believe that wetting was a serious problem under the conditions of these experiments.

The HREELS experiments were conducted without *in situ* XPS, so a direct thickness measure for each lubricant film was not possible. However, heating a lubricant covered surface in the XPS above the evaporator temperature resulted in films having thicknesses below 10 Å. This provided a nominal estimate of lubricant coverage in the HREELS experiments after heating to 473 K.

Friction measurements in UHV. Friction measurements were conducted at room temperature using a novel pin on flat device (*12, 13*) (fabricated by McAllister Technical Services, Coeur d'Alene, Idaho) attached directly to the preparation chamber in UHV. A photograph of the friction device is shown in Figure 2. This device conveniently allows for both pin and flat interchange in UHV by using a 'wobble' stick to remove specimens attached to VG stubs from either the deflection bar or collet and then place them on the transfer drive. The surfaces are then examined using XPS and / or AES in the same UHV system. The entire assembly is conveniently attached to a rotary platform (VG) mounted on an XYZ manipulator. In order to make a friction measurement, the deflection bar would then be rotated 180 ° from the configuration shown in Figure 2. The pin is secured by the cable clamp to prevent it from falling off the deflection bar. The flat is held rigid with a collet mechanism mounted on a rotary feedthrough. The load (0.10-0.20 N) was applied by translating the deflection bar along the load force direction until the desired load was attained. The pin was moved linearly 3×10^{-3} m at a speed of $\sim 6 \times 10^{-4}$ m/sec perpendicular to the applied load direction using a stepper motor coupled to the manipulator. Wear scars large enough for XPS analysis (0.003 x 0.003 m) were made by making ~100 juxtaposed (~30 μm apart) linear scars. Each line was made by performing two complete back and forth cycles. A Mac II computer (Apple Computer, Inc., Cupertino, California) was used to digitally record the load, adhesion, and coefficient of friction as a function of time with programs written using an icon based software system (LabVIEW software, National Instruments Co., Austin, TX). The wear scars were then studied using XPS in the same UHV system without exposure to air.

The pins and flats were machined from M50 steel rods (CarTech Company, Reading Pennsylvania). The manufacturer specified the following composition (atomic %): Fe-87.8%, Cr-4.4%, C-3.8%, Mo-2.4%, and V-1.1%. The M50 steel flats were cylinders having a diameter of 0.010 m and a thickness of 0.002 m. They were polished using SiC paper and alumina to a 0.3 μm finish. The M50 pins were machined and lapped to a 0.005 m radius and a 1 μm finish (Eureka Tool, Inc., Bridgeport, MI). All M50 surfaces were ultrasonically cleaned in acetone, followed by rinsing in methanol. Most M50 surfaces used in the XPS and friction experiments were further cleaned by exposure to ozone (*23*) using a commercially available ultraviolet / ozone cleaner (model T10X10 / OES, UVOCS, Inc., Montgomeryville, Pennsylvania) prior to insertion into UHV. Typical UV/ozone treatments lasted 600-900 sec and resulted in a 50-70% reduction of C on the surface compared to solvent cleaning.

Results

Thermal Surface Chemistry of HMPCP on M50 Steel Surfaces.
HMPCP / M50 Steel Interfaces. Figure 3 compares the HREELS spectrum to IR and Raman spectra of the liquid. The former was a thick (several layers deep) film of HMPCP adsorbed at 90 K on an M50 steel surface. The sample was thick enough to

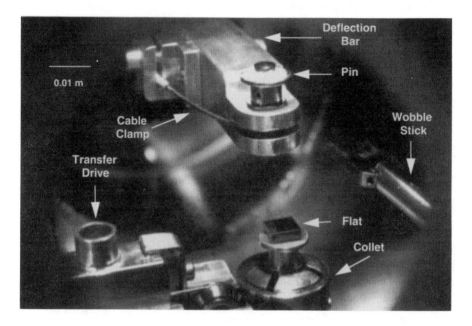

Figure 2. A photograph through the prep chamber view port showing the working parts of the friction device in UHV. A scale of 0.010 m is shown. The device allows for simple specimen interchange in UHV. See text for a discussion of the method of operating the device and how friction is actually measured.

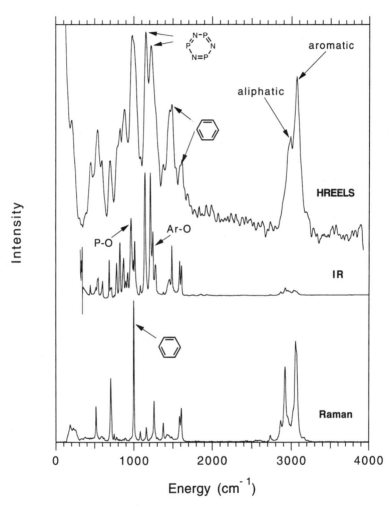

Figure 3. The HREELS spectrum of a thick layer of HMPCP on M50 steel measured at 90 K, the IR absorption spectrum, and the Raman scattering spectrum of the liquid, both at 298 K.

exclude variations in relative intensity caused by orientation or reaction at the surface. The vibrational spectrum measured by HREELS compares well with the IR spectrum except for differences in relative intensity of the CH stretching modes near 3000 cm^{-1}. The CH stretching modes are very strong in the HREELS. Several of the intense bands in the spectra have been labelled according to assignments of approximate functional groups which can be identified by their presence. Three bands at 1610, 1585 and 1487 cm^{-1} are attributable to bend stretches of the phenyl rings (24). Two bands which are very strong in both HREELS and IR at 1145 and 1211 cm^{-1} can be assigned to the phosphazene ring (25, 26). In the IR spectrum, the P-O-aryl moiety gives rise to two bands at 966 cm^{-1} (P-O stretch) and at 1245 cm^{-1} (O-aryl stretch) (24). Neither of these bands is fully resolved in the HREELS. The 1245 cm^{-1} O-aryl mode is unresolved from the 1211 cm^{-1} P-N mode. The P-O mode at 966 cm^{-1} appears to be convoluted with the band at 1000 cm^{-1} in Raman which is absent in IR. The latter can be identified as the phenyl ring breathing mode and is very often the most intense band in Raman but disallowed for IR absorption due to symmetry rules (27). The presence of the latter and the large CH stretching suggests that non-dipole scattering processes are important in the inelastic electron scattering cross sections here (28).

The evolution of the HREELS spectra of HMPCP on clean M50 as a function of temperature is shown in Figure 4a-d. The spectrum recorded at 90 K is the same as that in Figure 3. Heating to 473 K should have caused the desorption of any multilayer HMPCP. The two bands attributed to the phosphazene ring (Figure 4c) and that due to P-O (966 cm^{-1}) have decreased relative to the CH stretching modes after the 473 K heat. One of the phosphazene ring bands (1211 cm^{-1}) is reduced much more than the other (1145 cm^{-1}). The ratio of the intensities of aliphatic (~2950 cm^{-1}) to the aromatic CH stretching bands (~3050 cm^{-1}) was unchanged upon heating. This suggests that no decomposition of the phenyl ring or the methyl groups occurred (Figure 4c and 4d).

When this surface was heated to 750 K (Figure 4d), the low HREELS intensity in all losses suggests most of the adsorbed species have desorbed. The broad loss near 580 cm^{-1} is indicative of the formation of metal oxide. The loss near 1000 cm^{-1} may indicate the presence of some phosphate containing species. A small peak with a low energy shoulder (near 3000 cm^{-1}) attributable to CH stretches is still present after the 750 K anneal. Again, the ratio of aliphatic / aromatic CH stretching intensities are unchanged from the 90 K spectrum, consistent with no reaction occurring at the aromatic ring or scission of the methyl group from the ring. Because the loss labelled aryl-O in IR in Figure 3 is not resolved in HREELS it is difficult to follow the effect of heating on the aryl-O bond. It appears, however, there is some decrease in intensity in this mode and in the 966 cm^{-1} (P-O + phenyl breathing) after heating to 473 K. These observations are consistent with the decrease in C/O ratios and formation of metal oxide observed by XPS and discussed below.

The persistence of the 1145 cm^{-1} mode after heating indicates that P=N bonding remains. Two possible explanations for the decrease in intensity of the second P=N mode are phosphazene ring breaking or molecular orientation. The IR spectra of several model phosphazenes (25, 26) show no discernable dependence on cyclization in the P=N stretch region. This leads one to rule out PN ring breaking as the cause of the dramatic change in the relative intensities of the phosphazene ring modes. Orientation of the PN ring in HMPCP in the plane of the metal surface might also give rise to preferential strengthening of one mode over the other due to symmetry selection rules. It is difficult to assign the symmetry species of the two P-N ring modes in HMPCP (26), but it seems likely that the two modes assigned in Figure 3 to P=N have different symmetries and would therefore show different dependences on orientation. At the present time the orientation of the phosphazene rings relative to the metal surface is the most plausible explanation for the changes in the phosphazene modes.

XPS also provides insight into the thermal mechanism of HMPCP degradation. The stoichiometry of HMPCP using XPS is in agreement with the formula represented in

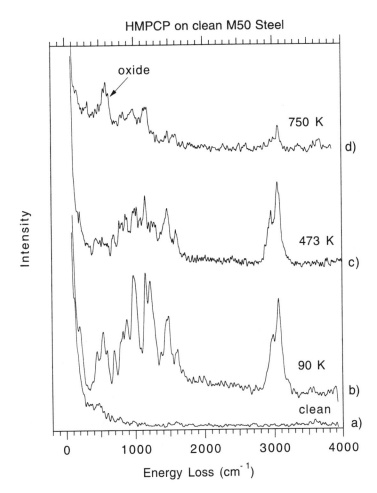

Figure 4. HREELS spectra of a) a clean M50 steel surface, b) the multilayer HMPCP adsorbed at 90 K, c) the surface in b) heated to 473 K, and d) heated to 750 K. The spectra are not normalized to the elastic intensities.

Figure 1. Figure 5 shows the O (1s) XPS spectra for HMPCP on clean M50 steel surfaces under various conditions. Figures 5b and 5c will be discussed in the mechanical interaction section. Figure 5a) shows that for a thick layer of HMPCP that has been shifted 0.7 eV to compensate for charging during the measurement. The single feature in Figure 5a) at 534.1 eV is representative of O bound to both P and organic C. Following heating to 400 K, the spectrum (solid line) shown in Figure 5d) was measured. The resultant fluid thickness is ~ 8 Å. The dotted trace represents the scaled background O present prior to HMPCP adsorption and the dot-dash trace is the resultant following subtraction of the background. From the dot-dash spectrum the formation of metal oxide species upon heating is evident. Based upon the binding energy the formation of a metal phosphate species cannot be ruled out. The ratio of O/C increases significantly upon heating to 500 K (~ 50 % increase). These results, along with the decreased intensities of the P-O and aryl-O modes in HREELS, suggest breaking of C-O bonds occurs. The relative concentration of N and P did not change upon heating. This may be due merely to the inability to measure changes in P and N due to the low sensitivity in XPS and low concentration in the bulk fluid.

Figure 6 shows the results of heating HMPCP on an oxidized surface. The bottom spectrum (Figure 6a) was recorded at 90 K after exposing the surface to 300 Langmuirs (1 Langmuir or 1 L = 1 x 10^{-6} Torr·sec) of O_2 at 295 K. The most intense loss (650 cm^{-1}) is due to the presence of metal oxide. Losses at 420, 1000, and 3650 cm^{-1} are attributed to a small amount of H_2O adsorbed from the residual gas in the UHV chamber. The 90 K spectrum (Figure 6b) was measured after adsorbing a thick layer of HMPCP onto the M50 surface. The layer of HMPCP adsorbed at 90 K was sufficiently thick to mask the vibrations from the oxide. This also means the layer was thick enough to mask vibrations due to adsorbates at the metal oxide - lubricant interface. The spectrum at 90 K (Figure 6b) can only be interpreted in terms of multilayer or condensed phase species. This spectrum is identical to that measured on the lubricant covered, clean surface at 90 K (Figure 4b). Heating HMPCP adsorbed on the oxidized surface to 473 K (Figure 6c) gives rise to changes in the HREELS spectrum similar to that observed for the clean surface. The 1211 cm^{-1} band decreases substantially and the intensity of the 966 cm^{-1} P-O stretching mode decreases. As seen for the clean surface, no evidence of changes in the aromatic / aliphatic CH stretch ratio are evident. A broad loss near 1000 cm^{-1} may indicate formation of a phosphate species on the surface.

HTFMPCP / M50 steel interfaces. A comparison of the HREELS spectrum of a thick layer of HTFMPCP to the IR and Raman spectra of the liquid is shown in Figure 7. The correlation of the three spectra is good, taking into consideration the differences in selection rules and resolution. Again, several of the vibrational bands are marked with approximate assignments of the functional groups which most strongly influence them. Losses attributable to the phosphazene ring stretching are unresolved from bands due to the CF3 group giving rise to a broad loss at about 1180 cm^{-1} with contributions from both groups. The band at 1320 cm^{-1} however should be fairly well resolved from interferences and should be able to serve as an indicator for the presence of CF3 moieties on the surface. An important observation in the HREELS spectrum is the absence of any CH stretching loss attributable to aliphatic CH.

Figure 8 shows an annealing set for a layer of HTFMPCP on the clean surface. The 'clean' surface (Figure 8a) was contaminated with a small amount of H_2O (librations near 500 cm^{-1}, HOH bend near 1600 cm^{-1}, and OH stretch at 3600 cm^{-1}) adsorbed from the residual gases in the chamber. Even at 90 K there is evidence of some reaction of the phenyl ring with the M50 steel surface clearly evident by the appearance of a loss peak at 2950 cm^{-1} which is due to aliphatic CH stretching. The presence of interfacial vibrations attributable to aliphatic CH stretching and losses attributable to water in the 90 K spectrum (Figure 8b) indicate that the HTFMPCP layer was relatively thin, probably less than 1-2 monolayers. Other features of the spectra of thin layers of HTFMPCP is

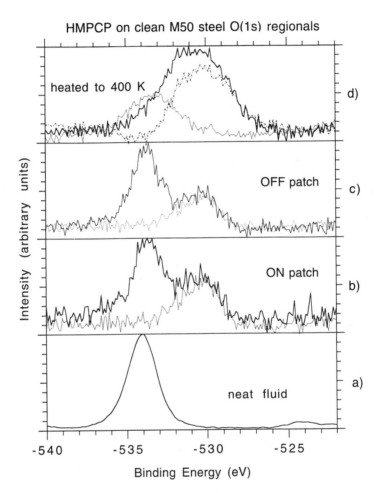

Figure 5. XPS O(1s) spectra of a) multilayer HMPCP on M50 steel compared to d) a spectrum measured for thinner (< 15 Å) layers after heating to 400 K d). The dotted line spectra are the background O (1s) spectra measured prior to dosing. The dash-dot curve in d) is a background subtracted spectrum showing metal oxide formation following heating to 400 K. The ON and OFF patch designations b) and c) are discussed in part B of the Results section.

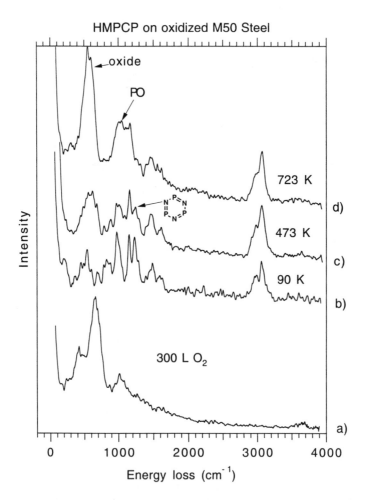

Figure 6. HREELS spectra of a) oxidized M50 steel, b) multilayer HMPCP adsorbed on oxidized M50 steel at 90 K, c) after heating the surface in b) to 473 K and d) 723 K.

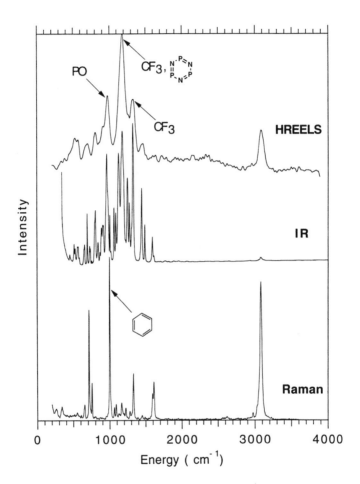

Figure 7. The HREELS spectrum of a thick layer of HTFMPCP on M50 steel measured at 90 K, the IR absorption, and the Raman scattering spectra of the liquid.

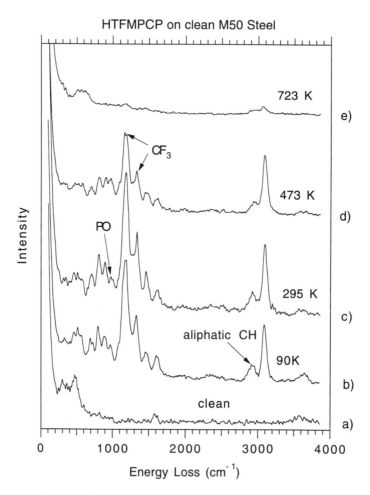

Figure 8. HREELS spectra of a) clean M50 steel, b) monolayer HTFMPCP adsorbed on M50 steel at 90 K and after heating to c) 295 K, d) 473 K, and e) 723 K. The spectra are not normalized to the elastic intensities.

the smaller intensities of the CF$_3$ and PO modes relative to CH stretching found in Figure 8b-e compared to that for HREELS spectra of multilayers of HTFMPCP (Figure 7). The decrease in intensity of the CF$_3$ modes suggest some reaction of the CF$_3$ groups occurs even at 90 K. Little change in the HREELS spectra occur for the sample heated 295 K (Figure 8c), but heating to 473 K (Figure 8d) continues the relative decrease in the intensity of the CF$_3$ modes relative to the CH stretch modes. The 723 K spectrum (Figure 8e) has a broad loss centered near 500 cm^{-1} which could either be due to metal oxide, metal fluoride or both left on the surface.

Further evidence for reaction at 90 K is obtained by examining the CH stretching region for both thin and multilayer HTFMPCP covered surfaces. The aliphatic CH stretching feature is absent for thick layers of HTFMPCP (Figures 7 and 9). When more fluid is added to low coverages of HTFMPCP the aliphatic stretching is masked as shown in Figure 9a. When a multilayer coverage of HTFMPCP is heated above the HTFMPCP dosing temperature (385 K), the aliphatic CH stretch loss reappears (Figure 9b). These experiments demonstrate that the aliphatic CH stretch loss is not due to an impurity in the fluid but is due to a genuine reaction product.

Figure 10 is a set of spectra for HTFMPCP on an oxidized M50 surface. The spectrum of the oxide, produced by heating at 295 K in 300 L O$_2$, consists of an intense broad loss near 600 cm^{-1} (Figure 10b). A thin layer of HTFMPCP was evaporated onto the oxide at 90 K (Figure 10c). The HREELS spectrum is similar to that for HTFMPCP deposited on the clean surface with the added feature due to the oxide. Reaction of the phenyl ring also occurs for adsorption on the oxide since at 90 K, a loss for aliphatic CH stretching (2950 cm^{-1}) is again present. Heating to 295 K caused no discernable changes in the HREELS spectrum. Heating to 473 K (Figure 10e) caused a decrease in the relative size of the aliphatic CH stretching suggesting desorption of the species responsible for the aliphatic CH stretch. Heating to 723 K (Figure 10f) causes desorption of much of the adsorbed species, but no further change in the aliphatic CH stretching. From these results, it appears that the oxidized surface is nearly or just as reactive as the clean M50 surface.

High resolution F (1s) XPS spectra for the clean and oxidized M50 steel surfaces are shown in Fig. 11. The 'ON and OFF patch' spectra (Figures 11b and e) will be discussed in the section on mechanical interaaction. Figure 11a shows the F (1s) spectrum for a thick layer of fluid. Only one chemical state is observed with a binding energy near 688.1 eV. The total F (1s) is ~ 25 at. % which agrees well with that in the fluid. HTFMPCP on clean M50 steel 20-25 Å thick at 298 K shows a small shoulder (up to 7 % of the total F (1s) peak area) towards lower binding energy (~ 684.2 eV) which can attributed to the formation of metal fluoride (Figure 11c). Upon heating to 510 K the metal fluoride feature at 684.2 eV increases in relative intensity (Figure 11d). The binding energy of this feature is consistent with the presence of metal fluoride (*29*). The metal fluoride feature comprises ~ 40 % of the total F present in the 5 Å thick film and the total F remaining is only 10 % compared with 30 % intensity for the unreacted fluid. This suggests that some F containing species desorb during heating and is consistent with the CF stretch intensity decrease in the HREELS. The spectrum after heating HTFMPCP on an oxide covered surface to 510 K is shown in Figure 11f. Similar reactivity and partitioning of F species is seen for the oxide compared to the clean metal surface. However, the metal-like F species is now shifted up to 684.9 eV binding energy. This may indicate that the species formed is not simply a metal F species. A similar F (1s) binding energy was reported for films formed in wear tests of steel surfaces interacting with PFPAE fluids under boundary lubricating conditions (*30*). A similar species was observed for 10-15 Å thick metal fluoride like films formed from reaction of perfluorodiethyl ether on clean Fe surfaces (*8*).

The C (1s) spectra for the identical series of surface treatments shown in Figure 11 are presented in Figures 12 a-f. Figure 12a shows the spectrum for a thick HTFMPCP layer. Three distinct C (1s) species were observed corresponding to C-C (285.0 eV), C-

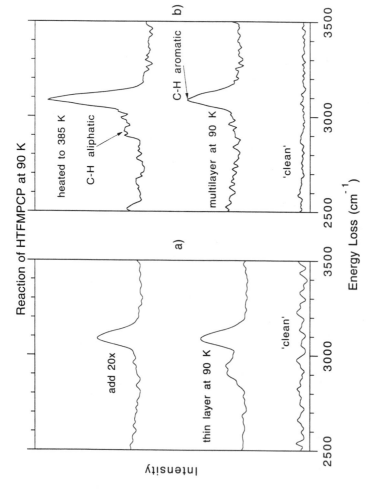

Figure 9. HREELS spectra measured at 90 K a) for a thin layer (~ monolayer) of HTFMPCP adsorbed on M50 steel (bottom) followed by ca. 20x larger dose (top). The aliphatic CH species disappears since it has been covered up by the additional HTFMPCP on the surface. In b) the CH stretching region is shown for a multilayer exposure of HTFMPCP at 90 K (bottom) and following heating to 385 K (top). The spectra are not normalized to the elastic intensities.

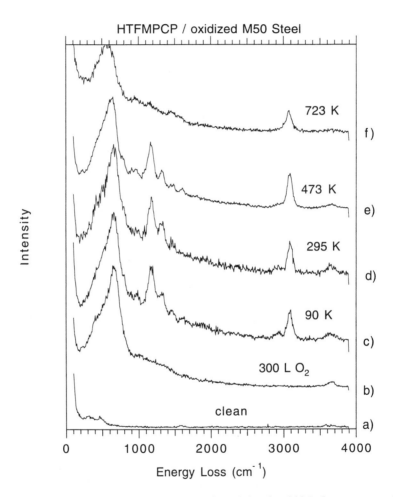

Figure 10. HREELS spectra of a) clean M50 steel, b) after 300 L O_2 exposure, c) nearly monolayer coverage of HTFMPCP adsorbed on oxidized M50 at 90 K and after heating to d) 295 K, e) 473 K, and f) 723 K.

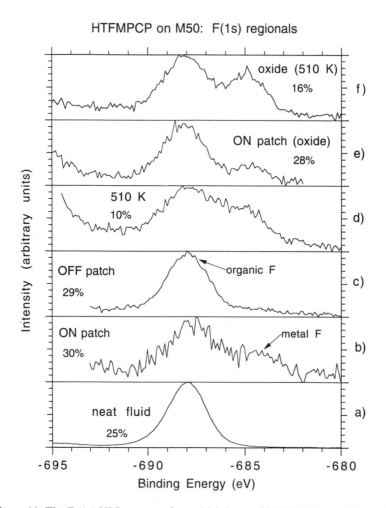

Figure 11. The F (1s) XPS spectra of a) a thick layer of HTFMPCP, b) ON patch after mechanical contact, c) adsorbed on clean M50 steel at 298 K (OFF patch), d) after heating the clean HTFMPCP covered surface to 510 K, e) ON patch after mechanical contact for an oxidized M50 steel surface, and f) after heating the oxidized HTFMPCP covered surface to 510 K. The HTFMPCP coverage is less than 15 Å in all cases. The ON and OFF patch designations (b, c, and e) are discussed in part B of the Results section. See text for further explanation.

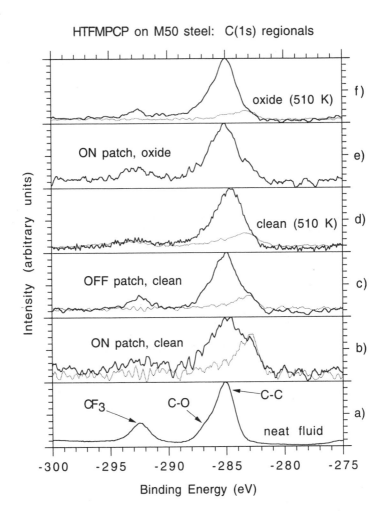

Figure 12. The corresponding C (1s) XPS spectra for the series shown in Figure 11 for the F (1s) region for a) a thick layer of HTFMPCP, b) ON patch after mechanical contact, c) adsorbed on clean M50 steel at 298 K (OFF patch), d) after heating the clean HTFMPCP covered surface to 510 K, e) ON patch after mechanical contact for an oxidized M50 steel surface, and f) after heating the oxidized HTFMPCP covered surface to 510 K. The HTFMPCP coverage is less than 15 Å in all cases. The ON and OFF patch designations (b, c, and e) are discussed in part B of the Results section. See text for further explanation.

O (286.5 eV), and C-F$_3$ (293.2 eV). Based on similar spectra recorded for HMPCP in the C (1s) region approximately 20-25 % of the intensity near 293.2 eV can be attributed to the $\pi\rightarrow\pi^*$ shake-up feature. This overlap makes it difficult to monitor changes in the aromatic portion of the fluid molecule due to reaction or complexing. The result of dosing the clean surface at 298 K is shown in Figure 12c. The dotted curve in Figure 12c is the scaled contribution due to the background carbon on the M50 steel surface. The binding energy of this carbon species (283.0 eV) is consistent with the presence of metal carbide present as an alloying agent in the steel. The spectra in Figure 12c indicate that no reaction of the fluid has occured. Upon heating the clean surface to 510 K changes in the C (1s) spectrum are evident in Figure 12d. The most dramatic change in the C (1s) spectrum compared with the unreacted fluid shown in (Figure 12a) is the decrease in the CF$_3$ functionality compared with other carbon species. This observation coupled with the decrease in F (1s) overall intensity points to the loss of CF$_3$ in the fluid upon heating the surface. Similar changes in the C-F stretching mode in the HREELS data are also consistent with this notion. Similar heating cycles conducting using the oxidized surface show a loss of the CF$_3$ functionality in XPS compared to the unreacted fluid. Although not shown, O (1s) spectra were also recorded before and after heating clean and oxidized surfaces covered with a thin layer of HTFMPCP. In both cases a new species was formed upon heating that had a binding energy of 531.5 eV. This is consistent with the formation of a small amount of an oxy fluoride species.

PPE / M50 steel interfaces. The interaction of PPE on M50 steel surfaces has been reported by DeKoven and Mitchell (*10*). The key features are summarized here for comparison with the phosphazene fluids. Figure 13 shows HREELS spectra for a thick layer of PPE adsorbed at 90 K (Figure 13a) and after heating to 420 K (Figure 13b). The most dramatic change which occurred here was the appearance of the aliphatic CH stretching (2950 cm^{-1}) and the relative decrease in the size of the C-O-C stretching modes (1200 cm^{-1}) after heating (Figure 13b). The appearance of the aliphatic CH stretching indicates electronic rehybridization (reaction) of the C atoms of the phenyl rings with the metal surface has occurred. The decrease in C-O-C stretch intensity suggests breaking of the C-O bonds have occurred and XPS results confirmed the formation of metal oxide-like species upon heating to 500 K. In addition, the C/O ratio decreases 50 % upon heating to the same temperature. Similar reactivity was found for PPE with the oxidized M50 surfaces.

Lubricant / M50 steel interfaces formed by mechanical interaction. Before showing the results of attempts to mechanically induce reactions it is useful to consider the typical friction behavior observed for the lubricant covered surfaces under the conditions of the measurements. It is important to emphasize that the primary motivation of these experiments is to observe the interfaces between metal surfaces and lubricants. In order to observe the interfaces using surface probes, thin layers of lubricant (≤ 30 Å) must be used. If thicker layers of fluid are used then the interfacial contributions will be masked. Typical friction traces for a lubricant covered surface are shown in Figure 14 for HTFMPCP during the process of making the 0.003 x 0.003 m scar. The behavior shown in Figure 14 is typical for all fluids examined. The M50 surface is covered with ~ 15 Å of HTFMPCP. Each of the 100 individual lines comprising the scar is made by two complete back and forth cycles of the pin across the flat. In all cases (Figure 14a-c) the initial pass of the pin across showed a lower friction (0.5-1.0) than subsequent traversals where the friction, μ, increases to 1.0-1.5. Stick-slip behavior is evident in all of the traces in Figure 14. This behavior is manifested by a slow rise in friction followed by a rapid drop. The large degree of stick-slip behavior present in the friction traces is consistent with the boundary mode of lubrication wherein the lubricant film is periodically broken during the sliding (*3*). This produces metal-metal and metal-oxide adhesion at those locations. The 15 Å layer of HTFMPCP provided significantly lower

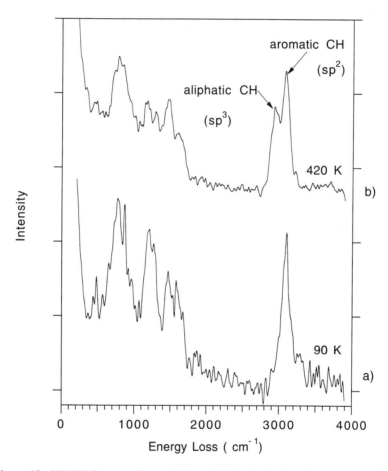

Figure 13. HREELS spectra for multilayer PPE adsorbed on clean M50 steel at a) 90 K and after heating to b) 420 K.

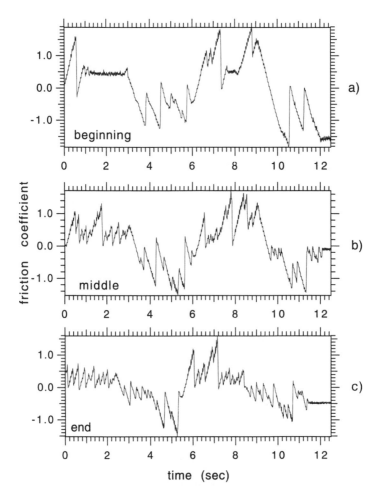

Figure 14. Typical friction traces during the fabrication of a 0.003 m x 0.003 m patch for two contacting oxidized M50 steel surfaces covered with 15 Å of HTFMPCP. The designations a), b), and c) each represent two complete back and forth cycles of the pin across the flat near the beginning, middle, and end of the patch, respectively. See text for further discussion.

friction than that measured for either the clean surface (μ =2.0-3.5) or for a 15 Å thick oxide film (μ=1.5-1.7). This is still much higher friction than that which is found for thicker ($\geq$ 100 Å) films of these lubricants (μ=0.1-0.2). For lubricant coverages $\geq$ 100 Å, it would be impossible to examine the metal / lubricant interface using XPS or HREELS.

Figure 15 shows the O (1s) spectrum for a PPE covered clean M50 steel surface which has undergone mechanical action to make a scar as described above. The solid line spectrum represents the M50 surface after tribological interaction using a thin PPE film. There are clearly two components in the spectrum. The higher binding energy feature (534.5 eV) is that due to O in PPE while the lower binding energy feature (530.5 eV) is consistent with the presence of metal oxide. The dotted spectrum is representative of the background oxide present on the M50 surface prior to lubricant deposition. The background oxide thickness is $\leq$ 0.3 Å (*10*) and could not be reduced by further sputter / heating cleaning cycles. Subtracting the O background (dotted spectrum) from the solid line spectrum gives the dot-dashed spectrum which clearly shows the enhanced formation of metal oxide due to tribochemical action. Although not shown, a much smaller amount of metal oxide is formed at 298 K in the absence of mechanical contact (*10*).

Scars made using the model phosphazene fluids were examined using XPS. No conclusive evidence for mechanically induced chemistry was observed for HMPCP. This is the result of several scars made using 20-30 Å thick HMPCP films in an identical fashion to those made using PPE. Although heating to 500 K results in metal oxide formation, no other chemical changes were observed using HMPCP on clean or oxidized M50 steel surfaces.

Evidence for a tribochemical induced reaction was observed for both clean and oxidized M50 steel surfaces using HTFMPCP as the lubricant film. This is illustrated by examining the F (1s) spectra shown in Figures 11b and 11e. Clearly, there is evidence for enhanced metal F formation on both surfaces compared to reaction at 298 K. Compared to the thermally induced reactions at 510 K there are some differences. First, the total F in the tribologically induced reactions is nearly equal to that for the unreacted fluid. This is in direct contrast to the dramatic reduction in total F for the thermally reacted layers. Second, the partitioning of the F species is quite different when heating to 510 K contrasted with tribochemical reactions. The metal F component is significantly larger for the thermally induced reaction.

Discussion

Properties of the phosphazene fluids and PPE which may influence their reactivity are activation of the phenoxy ring by substituents on the ring and steric constraints preventing the phenoxy ring from approaching the solid surface in the correct orientation to react. The trifluoromethyl, CF_3, group on the phenoxy ring in HTFMPCP is a strong electron withdrawing group and a good leaving group, which should activate the ring for nucleophilic attack relative to the methyl group in the HMPCP molecule (*31*). The reactivity of the CF_3 group is born out by the formation of metal fluoride detected by XPS, the decrease in the CF_3 stretching losses in HREELS, and the loss of F upon heating found by XPS. Reactions at the CF_3 group may facilitate further reaction or complexation of the phenoxy ring with the surface. The growth of sp^3 (aliphatic) type carbon detected by HREELS for PPE and HTFMPCP can be explained by complexation of the C atoms of the phenoxy ring with the solid surface. Complexation could be accomplished by forming at least two new sigma bonds between the surface and two carbon atoms of the phenyl ring to convert trigonally coordinated C atoms to tetrahedral coordination. It is interesting that HMPCP does not show phenoxy ring complexation similar to that observed for PPE and HTFMPCP. It may be that due to steric factors (*32*) when the HMPCP molecule lies on the surface, specific sites are not as accessible compared to PPE which could presumably lie flat. The ring must be properly oriented

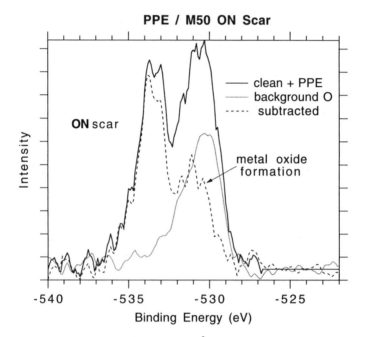

Figure 15. The O (1s) XPS spectra of a 20 Å PPE layer on clean M50 following mechanical contact. The dotted spectra represents the background O spectra due to residual oxide present prior to PPE dosing. The dash-dot curve is the resultant background subtracted spectrum showing formation of metal oxide due to mechanical contact.

with respect to the surface to allow the interaction to occur. PPE has no out of plane constituents on the phenyl rings which would prohibit the rings from lying flat on the surface. Additionally, the O atoms attached to the rings provide some activation of the rings and reaction with the surface appears to occur readily. In HMPCP and HTFMPCP, the orientation of the phenoxy rings with respect to each other is determined largely by interactions caused by bonding to the phosphazene ring. In the vapor phase, the rings lie in intersecting planes (*32*). It may be possible for the rings to reorient into a planar configuration due to interaction on a surface, but this is not sufficient for reaction of HMPCP with the surface. For both the cyclophosphazenes, if no reorientation occurs, the O end of the phenoxy ring will be (assuming a flat surface) too far from the surface to interact chemically. For HMPCP only a small amount of metal oxide is detected in XPS after heating to 500 K, but HTFMPCP and PPE reacted to form oxides or fluorides as well as tetrahedral C atoms. The reaction to form tetrahedral C appears to require reaction at one of the ring atoms activated by a substituent. In the case of HMPCP, the methyl group does not activate the ring and, along with the conformation caused by attachment to the phosphazene ring, prevents the interaction of the phenoxy O atom with the surface. HTFMPCP reacts at the F-C or $C-CF_3$ bonds and subsequently with the surface to form tetrahedral C and metal fluoride. PPE reacts at the aryl-O bonds and forms tetrahedral C and metal oxide.

The extent of reactivity detected for the oxide and the clean metal is quite similar. Both clean and oxidized metal surfaces can show either Lewis acid or base character depending upon the nature of the specific surface electronic interactions (*33*). At the present time it is difficult to say how the metal surface operates during reaction with PPE or the phosphazene fluids. For example in HTFMPCP the CF_3 would activate the phenyl ring towards nucleophilic attack. However, the CF_3 would also be susceptible to acid attack at the F atom sites. At the present time our data does not allow us to distinguish which site is reacted first. Models of site specific reactivity of fluorinated ethers are currently being developed (*34*). Other complicating factors are related to the actual coordination of the oxide surfaces and the distribution of oxide on the M50 surface. The question of surface impurities such as adsorbed water leading to OH formation and segregation of bulk impurities to the surface such metal phosphides and carbides upon heating could also have an influence on the reactivity.

A model for predicting the temperature of sliding surfaces has recently been presented by Lim and Ashby (*35*). They have presented a model that permits the estimate of bulk and flash temperature rises due to friction from the normalized pressure on the sliding interface, $\tilde{F}$, the normalized velocity, $\tilde{U}$, and friction. Physically, $\tilde{F}$ is the nominal pressure divided by the surface hardness and $\tilde{U}$ is the sliding velocity divided by the velocity of the heat flow. Bulk material properties were estimated using the values presented by Lim and Ashby for 'steel'. Using a contact area of 3×10^{-10} m^2 based on the measured diameter of a single line scar width and a load of 0.15 N, $\tilde{F}$ is 0.5. Similarly, $\tilde{U}$ is 1×10^{-3} using a sliding speed of 1×10^{-3} m/sec. Using μ of 2.0, the bulk temperature rise of the interface is estimated to be 0.1 K using Equation 8 of Reference 35. To estimate the flash heating during sliding Ashby and Lim postulate that the average area of an asperity contact is constant and the number, not the size, changes with load. According to Ashby and Lim, at equilibrium, the ratio of the nominal to real contact area is then simply equal to $\tilde{F}$. Using these assumptions, a flash temperature of 0.7 K was calculated for our conditions using Equation 17 in Reference 35. If the sliding speed during the stick-slip transition (~ ten times the normal sliding speed) is used, then a flash temperature of 7 K is calculated. Based on these estimates of flash heating, it is unlikely that the tribochemically induced reactions are caused merely by thermochemical processes.

There are at least three other possible mechanisms which could explain the tribochemistry observed for PPE and HTFMPCP. The first mechanism involves enhanced surface cleaning, i.e. removal of species covering the metal, during rubbing. When two oxidized Fe surfaces were rubbed together at 298 K, the surface cleanliness was found to exceed that obtained by sputter cleaning (*12*). This may make the contacted surfaces more reactive than surfaces cleaned in UHV by sputtering or heating. The influence of enhanced cleaning cannot be totally ruled out, but seems unlikely to be a significant factor since similar reactivities were found for both the oxide and clean M50 surfaces. The second mechanism, related to the first, involves producing types of coordinatively unsaturated surface atoms due to contact which may not be present on uncontacted surfaces. These sites may be more reactive than those on normally cleaned surfaces. The third and most probable mechanism is direct shearing of molecular bonds in the fluid molecule comprising the boundary layer, mechanochemistry. Induced shear to the boundary layer is a result of the relative motion of the sliding surfaces. If only one layer of fluid molecules separates the two sliding surfaces, all the shear stress except that dissipated by plastic deformation of the solid surfaces must be concentrated in that single monolayer. If the molecule is anchored strongly to the surface (PPE and HTFMPCP), then some of this stress may be relieved by breaking of bonds in the fluid molecule. In contrast, for the case of weak bonding to the surface (HMPCP), stress may be relieved by plowing through the lubricant film. This action displaces the fluid molecules, breaking the surface-fluid molecule bond.

Summary and Conclusions

Despite the similarity of the two phosphazene molecules, differing only by fluorination of the methyl group on the phenoxy substituents, a substantial difference in the surface and tribochemistry was observed. We have presented evidence for tribochemistry of HTFMPCP similar to that observed for PPE on M50 steel surfaces (*10*). In contrast, no evidence for tribochemistry was seen for HMPCP under similar conditions. Thermally induced chemistry was demonstrated on the 'clean' and oxide covered surfaces for both phosphazenes after heating to > 470 K. The HTFMPCP fluid reacted both at the aromatic rings, as did PPE, and at the CF_3 group. Surprisingly, HMPCP showed no reactivity of the aromatic rings. Breaking of the aryl-O bond was observed for both HTFMPCP and HMPCP. Chemisorption of HMPCP also led to changes in the HREELS spectra which are interpreted in terms of preferred orientation of the phosphorous-nitrogen ring. The tribochemistry is not believed to be thermally induced due to the low loads and sliding speeds. The most likely mechanism for the observed tribochemistry is one in which the shear stress due to sliding causes mechanical bond breaking (mechanochemistry) in the HTFMPCP and PPE molecules.

Acknowledgments

The following individuals from the Dow Chemical are thanked for their contributions. M. McKelvy, C. Putzig, and R. Nyquist are thanked for measuring and interpreting the infrared spectra of the various lubricant molecules, C. Pawloski and D. Wilkening for providing the purified phosphazene lubricant molecules, C. Langhoff for the fluid index of refraction measurements, M. Keinath for obtaining ellipsometry data, N. Rondan for discussions about reaction mechanisms and providing information about the phosphazenes structures, J. Womack and D. Hawn for critically reading this manuscript, and R. Chrisman, W. Knox, and T. Morgan for their encouragement to conduct this research. In addition, Dr. S.V. Pepper of the NASA Lewis Research Center is also acknowledged for many helpful discussions regarding thin lubricant films and Professor Y-W. Chung of Northwestern University for discussions about surface temperatures during sliding.

Literature Cited

1. Bowden, F.P.; Tabor, D. *The Friction and Lubrication of Solids II*; Oxford University Press: London, U.K., 1964.
2. Buckley D.H. *Surface Effects in Adhesion, Friction, Wear, and Lubrication*; 1st edition; Elsevier Scientific Publishing Company: The Netherlands, 1981.
3. Rabinowicz E. *Friction and Wear of Materials*; John Wiley & Sons: New York, NY, 1965.
4. Walczak M.M.; P.A. Thiel *Sur. Sci.* **1989**, *224*, p. 425.
5. Leavitt P.K.; Thiel P.A. *J. Vac. Sci. Technol. A* **1990**, *8*, p. 2269.
6. V. Maurice V.; Takeuchi K.; Salmeron M.; Somorjai G.A. accepted in *Sur. Sci.* **1991**.
7. Napier, M.E.; Stair, P.C. *J. Vac. Sci. Technol. A* **1991**, *9*, p. 649.
8. DeKoven, B.M.; Meyers, G.F. *J. Vac. Sci. Technol. A* **1991**, *9*, p. 2570.
9. Herrera-Fierro, P.C.; Jones, Jr., W.R.; Pepper, S.V. "Interaction of Fluorinated Polyethers with Metal Surfaces Studied by XPS", to be published in *J. Vac. Sci. Technol. A* **1991**.
10. DeKoven, B.M.; Mitchell, G.E. accepted in *Appl. Sur. Sci.* **1991**.
11. Miyoshi, K.; *Mat. Res. Soc. Symp. Proc.* **1989**, *153*, p. 321; NASA TM-101959 **1989**.
12. DeKoven, B.M.; Hagans, P.L. *J. Vac. Sci. Technol. A* **1990**, *8*, p. 2393.
13. DeKoven, B.M.; Hagans, P.L. *Mat. Res. Soc. Symp. Proc.* **1989**, *140*, p. 357.
14. Nadar, B. S.; Kar, K. K.; Morgan, T. A.; Pawloski, C. E.; Dilling, W. L. "Development and Tribological Properties of New Cyclotriphosphazene High Temperature Lubricants for Aircraft Gas Turbine Engines," 46th STLE Annual Meeting(Preprint No. 91-AM-7D-1) to be published in *Tribology Transactions* **1991**.
15. Mahoney, C.L.; Barnum, E.R.; Kerlin, W.W.; Sax, K.J. *ASLE Trans.* **1960**, *3*, p. 88.
16. Jones, Jr., W.R.; *ASLE Trans.* **1985**, *29*, p. 276 and references therein.
17. Jones, Jr., W.R. "The Tribological Behavior of Polyphenyl Ether and Polyphenyl Aromatic Lubricants", *NASA TM-10016*, **1987** and references therein.
18. Chakrabarti, A., Dow Chemical Company, unpublished data from Knudsen effusion measurements for PPE and several different phenoxy substituted cylcotriphosphazenes.
19. Seah, M.P. in *Practical Surface Analysis*; Briggs, D., Ed.; Seah, M.P., Ed.; John Wiley & Sons: Chichester, UK, 1983; p. 181.
20. Azzam, R.M.A.; Bashara, N.M. *Ellipsometry and Polarized Light* ; North Holland: Amsterdam, 1987.
21. McCrackin, F.L. *NBS Technical Note 479*, 'A Fortran Program for Analysis of Ellipsometer Measurements' **1969**.
22. Novotny, V.J. *J. Chem. Phys.* **1990**, *92*, p. 3189.
23. Vig, J.R. *J. Vac. Sci. Technol. A* **1985**, *3*, p. 1027 and references therein.
24. Socrates, G. *Infrared Characteristic Group Frequencies*; John Wiley & Sons: Chichester, U.K., 1980.
25. Allcock, H.R. *Phophorous-Nitrogen Compounds*; Academic Press: New York, NY, 1972.
26. Allcock, H.R.; Kugel, R.L.; Valan, K.J. *Inorg. Chem.* **1966**, *5*, p. 1709.
27. Colthup, N.B.; Daly, L.H.; Wiberley, S.E. *Introduction to Infrared and Raman Spectroscopy*; Academic Press: New York, NY, 1964.
28. Ibach, H.; Mills, D.L. *Electron Energy Spectroscopy and Surface Vibrations*; Academic Press: New York, NY, 1982.
29. Wagner, C.D.; Riggs, W.M.; Davis, L.E.; Moulder, J.F.; Muilenberg, G.E. *Handbook of X-Ray Photoelectron Spectroscopy*; Perkin-Elmer Corporation: Eden Prairie, MN, 1979.

30. Carré, D.J. *ASLE Trans.* **1986**, *29*, p. 121.
31. Lowry, T.H.; Schueller Richardson, K. *Mechanism and Theory in Organic Chemistry*; Harper & Row: New York, NY, 1976; p. 395.
32. Rondan, N., Dow Chemical Company, private communication and by constructing a 3-D space filling model of the phosphazenes.
33. Stair, P.C. *J. Am. Chem. Soc.* **1982**, *104*, p. 4044.
34. Napier , M.E.; Stair, P.C., Northwestern University, private communication and to be published.
35. Lim, S.C.; Ashby, M.F. *Acta. Metall.* **1987**, *35*, p. 1 and references therein.

RECEIVED October 21, 1991

Chapter 3

The Influence of Steel Surface Chemistry on the Bonding of Lubricant Films

Stephen V. Didziulis, Michael R. Hilton, and Paul D. Fleischauer

Mechanics and Materials Technology Center, Aerospace Corporation, El Segundo, CA 90245

The surface chemical composition of 440C steel and changes in composition caused by chemical and physical treatments are studied with X-ray photoelectron spectroscopy (XPS). Both an XPS sputter profile and the angular dependence of XPS core levels of solvent cleaned 440C indicate the existence of an ~ 2.5-nm-thick, iron oxide overlayer and an ~ 1.5-nm-thick chromium oxide underlayer on top of the bulk steel. Treatments with an HCl/ethanol or a buffered alkaline solution remove the iron oxide overlayer, while ion sputtering removes the oxides in addition to reducing Fe_2O_3 to metallic Fe. Removal of the iron oxide overlayer enhances the adhesion of MoS_2 lubricant films. The oil additive lead naphthenate physisorbs on the iron oxide overlayer at room temperature but chemisorbs on the pretreated surfaces or when the oxide layers are physically removed.

The preponderance of mechanical assemblies manufactured from steels make the lubrication of steel component surfaces the predominant problem in the field of tribology. In certain situations, such as the use of solid lubricants and boundary additives in liquid lubricants, the steel surface chemistry must play an intimate role in determining the success or failure of the lubricant. In particular, understanding the adhesion of solid lubricant films to components and the performance of extreme pressure (EP) oil additives operating under boundary conditions requires knowledge concerning the chemical composition and reactivity of the steel surface under study. In this paper, the chemical composition of the surface region of AISI 440C steel was explored with X-ray photoelectron spectroscopy (XPS) in order to determine the roles of the various surface species in interactions with radio frequency (rf) sputtered molybdenum disulfide (MoS_2) solid lubricant films and with the EP oil additive lead naphthenate (Pbnp).

Steel surface chemistry has been the subject of substantial amounts of experimental work.(1) High chromium stainless steels (such as 440C) generally

0097–6156/92/0485–0043$06.00/0

show segregation of chromium to the surface region; a corrosion barrier is obtained once a chromium oxide layer has formed.(2) In this work, detailed XPS studies of the oxide structure of 440C are presented, along with data examining the effects of chemical treatments on the surface composition. Finally, the effects of changing the steel surface composition on the adhesion and growth of MoS_2 films and the bonding interactions of Pbnp are also reported.

The two lubricant materials used in this work, MoS_2 and Pbnp, have chemical bonding and reactivity properties which enhance their tribological performance. The lubricating properties of MoS_2 result from its anisotropic, layered structure.(3) The material forms sandwiches of S-Mo-S having very strong Mo-S bonds, but only weak van der Waals interactions exist between sulfur layers of adjacent sandwiches. The weak van der Waals interactions result in very low shear strength perpendicular to the MoS_2 c-axis and produce highly inert, S-terminated (0001) basal planes.(4) The adhesion and growth of sputter deposited MoS_2 films are also affected by this structure, with reactive edge plane atoms acting as nucleation sites for faster film growth.(5) Pbnp provides decreased wear and low friction for steel surfaces operating under EP conditions of metal-to-metal contact and high temperature. The method by which Pbnp provides protection is unknown, but interaction with the steel surface must take place, leading to adsorption and/or reaction on the steel, for boundary protection to occur. The principal surface analysis technique employed in this work is XPS, which gives both quantitative surface composition information from peak intensities and qualitative information regarding the oxidation state and bonding environments of surface species from the binding energy positions of peaks.

Experimental Section

Samples of AISI 440C steel were cut into coupons approximately 6 x 6 x 2 mm in size. The samples were polished with slurries of alumina in water down to 0.05 μm grit size, cleaned in ethanol, and stored in a nitrogen purged desiccator until needed. Before introduction into the surface analysis vacuum system or deposition of lubricant species, the steel samples were subjected to a variety of chemical treatments while being agitated in an ultrasonic cleaner. Some samples were simply cleaned once again with ethanol. Others were subjected to acid etches for 30 s in a 20% concentrated HCl/ethanol solution followed by several ethanol rinses. Additional samples were cleaned for 1 h in a commercial, buffered alkaline solution (Alum Etch 34), ~15% in deionized water (pH = 12.8), followed by a brief rinse with deionized water and ultrasonic cleaning in ethanol. All treatments were conducted in the laboratory ambient; exposure to air was limited by keeping samples immersed in ethanol following treatments.

Thin films of MoS_2 were deposited onto the steel substrates in an rf sputtering system, which has been described elsewhere.(6) The steel coupons were placed in the diffusion pumped deposition system following chemical treatment and pumped down overnight (base pressure 1 x 10^{-6} Torr). The 1 μm thick films were produced at a deposition rate of 27.5 nm/min, with the sample temperature floating and reaching a maximum of 70°C. MoS_2 film structure and adhesion were studied by fracturing the film with a diamond brale indenter (200 μm radius of curvature) in a Rockwell hardness tester (150 kg load), followed by analysis with scanning electron microscopy (SEM) in a Cambridge Stereoscan S-200. The details of this procedure are outlined elsewhere.(7)

The steel surfaces were exposed to Pbnp by immersing the differently pretreated coupons into a 1-2 g/L solution of Pbnp in heptane for 1 min. During immersion, areas of the samples (approximately 2 x 2 mm) were scratched with a diamond scribe with sufficient force to plastically deform the surfaces to a depth of 1 to 2 μm as determined by surface profilometry. The scratching procedure was pursued in order to expose bulk steel to the Pbnp solution by breaking through the thin oxide layers. Following immersion, the samples were rinsed with heptane and immersed in ethanol. Prior to introduction into the ultrahigh vacuum (UHV) system for XPS analysis, the samples were exposed to air for approximately 5 min during mounting.

The surface composition of the treated steel surfaces and the chemical reaction of Pbnp on steel were studied with a Surface Science Instruments (SSI) SSX-100 XPS system. The UHV system (base pressure < 1 x 10^{-10} Torr) and instrument have been described elsewhere.(8) The 300-μm spot size from the monochromatized Al Kα anode and 50-eV pass energy used for this study produce a Au 4f$_{7/2}$ peak at 84.0 eV with a full-width-at-half-maximum of 0.95 eV. All binding energies are referenced to the spectrometer Fermi level. Data were obtained with the sample normal tilted by 60° relative to the central axis of the electron energy analyzer unless otherwise stated. An XPS sputter profile of a solvent cleaned 440C surface was performed with 2 keV Ar$^+$ ions to study the oxide layer structure. Spectra were obtained on the unscratched and scratched regions of the Pbnp exposed samples. The Pbnp treated samples were also heated in the UHV system to simulate temperature increases expected during boundary contact.

XPS Results and Discussion

XPS Sputter Profile. Figure 1 presents two different methods for determining the oxide layer structure on solvent cleaned 440C. Figure 1(a) is an XPS sputter profile, giving atomic concentrations assuming a homogeneous sample as a function of sputter time. The profile shows four different regions, separated by dashed lines, to exist on the 440C sample. Before sputtering, a strong carbon peak having a binding energy near 285 eV is evident, along with the Fe, Cr, and O signals expected for an oxidized steel surface. Both the characteristic C 1s binding energy and its rapid removal with sputtering (<1 min) indicate that the surface is contaminated with a layer of hydrocarbons. From 1 to 6 min of sputter time, the O signal decreases linearly while the Fe intensity remains about constant and the Cr signal increases. During this sputtering period, the Fe 2p XPS peaks show a monotonic decrease in the signal associated with iron oxide species (primarily Fe$_2$O$_3$) relative to the metallic iron signal. In addition, both the chromium oxide and chromium metal peaks increase in intensity, maintaining the same relative oxide-to-metal peak intensities. This behavior is consistent with the removal and possible chemical reduction of a layer of iron oxides, leading to an increase in the intensities of underlying species (metallic iron, chromium oxides, and metallic chromium).(9) In the 6 to 9-10 min sputter time frame, the oxygen signal decreases sharply, while the iron signal increases sharply and chromium shows a slight decline. The Cr 2p XPS peaks show that chromium oxides (primarily Cr_2O_3) are being sputtered away, eventually leaving behind a surface that is quite similar in composition (72% Fe, 18% Cr, 8% C, 2% O) to the bulk steel. A diagram of the steel surface region is given above the sputter profile, including oxide thicknesses determined by the calibration of the Ar^+ ion etch rate (0.42 nm/min) with an SiO_2/Si sample of known oxide thickness.

The abrupt changes in the slopes of the surface composition profile and the peak shape changes observed in the Fe and Cr core levels with sputtering are a direct result of the layered oxide surface structure on 440C. If the surface region were composed of mixed iron and chromium oxides or gradually changing compositions (graded interface), then much more gradual changes would have been observed. In fact, it is difficult to support the existence of anything but a layered oxide structure with fairly sharp boundaries between the metal and chromium oxide layer and between the two oxide layers based on this data. Only a very small amount of iron oxide intensity is evident during the early stages (first minute) of sputtering through the chromium oxide underlayer. While this signal could be taken as proof of the existence of some interfacial mixing of the oxides, it could also be explained by surface roughness effects, by knock-on effects, or by the existence of a small amount of an interfacial iron-chromium oxide spinel compound which has been proposed to exist.(10)

Angle Resolved XPS. Figure 1(b) gives the Cr $2p_{3/2}$ and the Fe $2p_{3/2}$ core levels obtained as a function of sample angle relative to the central axis of the electron energy analyzer. Moving down the figure, the sample normal has been tilted away from the analyzer, decreasing the effective escape depth of the electrons from the material and increasing the surface sensitivity of the technique. This angular effect allows the relative positions and thicknesses of layers in a material to be determined without destroying the interface with ion sputtering. The data in Figure 1(b) are normalized to the intensities of the higher binding energy iron and chromium oxide features and show that rotating the sample away from the analyzer increases the intensities of both the iron and chromium oxides relative to the metal signals. This result proves that the oxides exist on top of the metallic species. In addition, the decreasing effective escape depth has a greater impact on the absolute chromium oxide signal intensities than on the iron oxide features as indicated by the greater normalization factors listed for the Cr spectra on the figure. This result shows that the chromium oxides are deeper into the bulk of the sample than the iron oxides, in agreement with the sputter profile.

The angular intensity effects can be quantified for comparison to the sputter profile results just discussed. Equations used to study the attenuation of substrate XPS peaks by an overlayer(11) can be readily modified for the duplex overlayers present on 440 C as shown in equation 1.

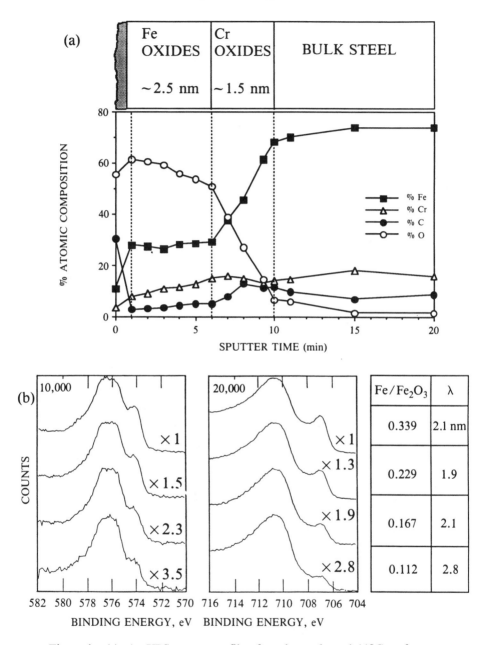

Figure 1. (a) An XPS sputter profile of a solvent cleaned 440C surface with an accompanying diagram showing the layered oxide structure of the surface region. (b) The Cr (left) and Fe $2p_{3/2}$ peaks as a function of sample angle ϕ from equation 1. From the top, the angles between the surface normal and the electron analyzer entrance cone are 0°, 20°, 40° and 60°. The table at the right gives the experimental Fe/Fe_2O_3 intensity ratios and the calculated electron escape depths.

$$\frac{I_{Fe}}{I_{Fe_2O_3}} = \frac{I^0_{Fe}e^{(\frac{-(x+y)}{\lambda cos\phi})}}{I^0_{Fe_2O_3}[1 - e^{(\frac{-x}{\lambda cos\phi})}]} \tag{1}$$

In equation 1, I_{Fe} and $I_{Fe_2O_3}$ are the integrated XPS intensities of the iron metal and iron oxide components of the Fe $2p_{3/2}$ spectrum, respectively; the I^0's are the intensities expected for single component systems; x (y) is the thickness of the iron (chromium) oxide overlayer; λ is the photoelectron scattering length in the oxide overlayer; and ϕ is the angle between the surface normal and the analyzer axis. The value of λ is assumed to be the same in both oxide layers. The I^0 ratio $[I^0(Fe)/I^0(Fe_2O_3)]$ is calculated to be 1.57 from the relative Fe atomic densities in 440C steel and Fe_2O_3, assuming the photoionization cross sections of the two iron species to be the same.

In order for equation 1 to be used, a value for λ must be determined for a Fe 2p photoelectron with a kinetic energy of ~780 eV. There are methods for calculating $\lambda(12)$, but their veracity is still a matter of conjecture, especially for a complex, multiphase material such as 440C. Instead, we will use the oxide thickness values determined by the sputter profile to calculate a value for λ from equation 1. (The oxide thicknesses determined by the sputter profile could be in error if the sputtering rates of the metal oxides are significantly different than that for the calibration sample.) The total oxide thickness (d) was determined to be 3.8-4.4 nm, with the iron oxide (x) 2.3-2.7 nm and the chromium oxide (y) 1.3-1.7 nm thick. Experimental peak intensities were determined by fitting the data with the SSI data analysis software using Gaussian broadened Lorentzian peaks and a Shirley background; the Fe/Fe_2O_3 integrated intensity ratios are given in Figure 1. The λ values obtained from equation 1 at the four different angles with x = 2.5 nm and y = 1.5 nm (also given next to the Fe spectra in Figure 1) are fairly consistent and result in an average value of 2.2 nm. This λ is reasonable for 780 eV electrons but slightly larger than expected when compared to values obtained from the "universal curve" of electron mean free paths.(13) The larger λ could be the result of inhomogeneous oxide thicknesses due to surface roughness effects as the calculated value of λ is greatest at the largest angle [Figure 1(b)].

Chemical Treatments of 440C. The effects of chemical treatments on the steel surface composition are shown in Figure 2, which compares the Fe and Cr $2p_{3/2}$ peaks following the specified treatment. The most significant changes are observed in the Fe data, which show a large decrease in the amount of surface iron oxides following both the acid etch and the alkaline wash. In contrast, the Cr peaks show little change in the oxide:metal relative intensities. The total Fe/total Cr peak intensity ratios decrease greatly from typical values of 3.2 for the solvent cleaned samples to values ranging from 1.3 to 2.0 for the acid and base treated steel. These results lead to a clear conclusion: the chemical treatments selectively remove the iron oxide overlayer, leaving behind a surface enriched in chromium

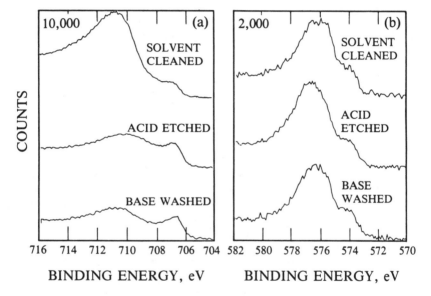

Figure 2. The (a) Fe $2p_{3/2}$ and (b) Cr $2p_{3/2}$ XPS peaks as a function of the sample treatments listed on the figure.

oxides. In addition, the chemical treatments are more effective at selectively removing the iron oxide layer than ion sputtering, where the smallest Fe/Cr ratio was 2.1 after the sixth minute of sputtering. This result likely confirms that ion bombardment chemically reduces some of the Fe_2O_3 to metallic iron in addition to removing material, causing a larger-than-expected value of Fe/Cr. Acid etched samples sometimes took on a milky appearance following the treatment, indicative of surface damage. If the samples were etched longer than 30 s, an increase in the amount of surface iron oxide was detected by XPS, showing that the acid had etched through the chromium oxide layer, initiating corrosion of the bulk steel.(9) Surface profilometry of acid etched surfaces indicated that samples etched for 0 to 30 s had average surface roughness of ~6 nm while a 60 s etch had an average roughness of ~17 nm.

The effectiveness of the chemical treatments can be investigated through intensity analyses of the Fe/Cr ratios. If all of the iron oxide overlayer has been removed, these intensity ratios should be defined by equation 2.

$$\frac{I_{Fe \, Metal}}{I_{Cr \, Total}} = \frac{I_{Fe}^0 e^{\left(\frac{-y}{\lambda_{Fe} \cos\phi}\right)}}{I_{Cr_2O_3}^0 \left[1 - e^{\left(\frac{-y}{\lambda_{Cr} \cos\phi}\right)}\right] + I_{Cr}^0 e^{\left(\frac{-y}{\lambda_{Cr} \cos\phi}\right)}} \tag{2}$$

Values for the I^0's must take into account the different metal atom densities in the various materials and the different Fe 2p and Cr 2p photoionization cross sections.(14) If some simplifying assumptions are made, setting $\lambda_{Fe} = \lambda_{Cr} = 2.2$ nm and $y = \lambda\cos\phi$, then a Fe/Cr ratio of 1.3 is obtained. This Fe/Cr value is close to the range obtained after the chemical treatments (1.3 to 2.0), lending credence to this approach. More accurately, the slightly larger λ_{Cr} (for photoelectrons having kinetic energies of 910 eV) and the greater thickness of the Cr_2O_3 layer (1.5 nm > $\lambda\cos\phi$ for $\phi = 60°$) would tend to lower this ratio. For example, if $\lambda_{Cr} = 2.5$ nm and $y = 1.5$ nm, then the calculated intensity ratio becomes 0.87. In our experiments, some iron oxide signal remains after all chemical treatments, resulting in Fe/Cr ratios in excess of this lower bound. The lowest experimental Fe/Cr ratios were obtained after alkaline washes (1.3 to 1.7), showing that this treatment was the most effective at selectively removing the iron oxide overlayer. Samples treated with the acid solution had a range of Fe/Cr ratios of 1.3 to 2.0. The acid etch data presented in Figure 2 shows the most selective iron oxide removal for any of the samples treated with this solution in this study (i.e., Fe/Cr = 1.3). As indicated above, most of the acid etched samples had higher Fe/Cr ratios than the alkaline washed samples.

The selective removal of iron oxide by the chemical treatments gives further support for the layered oxide structure indicated by the XPS results. The fact that some iron oxide signal remains after all sample treatments might support the existence of a small amount of an iron-chromium oxide compound that is not affected by the chemical treatments.

Interaction of 440C with Lubricant Films.   The effects of the chemical pretreatments on the adhesion of the rf-sputtered MoS_2 films deposited on 440 C are shown in Figure 3.   The SEM micrographs presented in this figure show the extent of delamination around the Rockwell indentations.   Qualitatively, it is obvious that the removal of the iron oxide overlayer by the acid and base treatments improves the fracture toughness and probably enhances the adhesion of the MoS_2 film.   Similar enhancements of adhesion are observed if the steel substrate is Ar^+ ion cleaned in the sputter system prior to film deposition.(9)   In addition, higher magnification micrographs indicate that the films deposited on the chemically treated surfaces appear slightly denser than on a solvent cleaned 440C surface.   In previous work, denser MoS_2 films of a given microstructure had a *greater* tendency to delaminate because of their reduced ability to blunt crack propagation due to decreased film porosity.(15)   Although these results do not conclusively indicate that the MoS_2 film has stronger bonding interactions with the chromium oxide components of the steel, the effects of the chemical pretreatments on film nucleation and adhesion indicate that more active sites for crystallite nucleation and strong adhesion exist on the steel surfaces after the chemical treatments.   This result implies that the effect is dominantly chemical in nature. An increase in surface roughness could conceivably cause an increase in adhesion based on mechanical interlocking effects, but the samples used in this work displayed little change in surface roughness following sample treatment.   In previous work, no enhancement in MoS_2 adhesion was observed for 440C surfaces etched for 60 s relative to a 30 s etch sample despite a large increase in the measured surface roughness.(9)   We conclude, therefore, that the enhanced adhesion is predominantly chemical in nature.

The final area of study to be discussed is the interaction of the EP additive Pbnp with the steel surfaces.   The Pbnp molecule consists of cyclopentane terminated hydrocarbon chains of varying length bonded to a central Pb ion through the oxygen atoms of a carboxylate group.   Figure 4 shows the Pb 4f and C 1s peaks obtained from Pbnp treated 440C surfaces as a function of sample pretreatment.   On the solvent cleaned surface (1), the Pb $4f_{7/2}$ peak has a binding energy of 139.2 eV, which has been identified as the binding energy for molecular Pbnp.(16)   The C 1s spectrum from this same surface is quite complex, with a small peak near 283 eV from steel carbides, strong features near 285 and 286 eV, binding energies characteristic of hydrocarbons both in the naphthenate chain and as impurities, and a peak near 289 eV.   The 289 eV binding energy is characteristic of the carboxylate carbon species expected for molecular Pbnp. Therefore, the strong carboxylate peak and the 139.2 eV Pb $4f_{7/2}$ binding energy show that Pbnp is molecularly adsorbed (physisorbed) on iron oxide.(17)

The Pb and C XPS peaks obtained from the scratched area of the solvent cleaned sample (2) as well as the unscratched areas of the acid (3) and base treated (4) 440C samples are also included on Figure 4.   For all of the altered surfaces, the Pb $4f_{7/2}$ peak is shifted to lower binding energy by 0.3 to 0.4 eV and the carboxylate C peak intensity is much weaker compared to solvent cleaned 440 C. These changes indicate a different chemical interaction of Pbnp with the steel has occurred.   The spectroscopic changes are consistent with partial breakdown of

500 μm   100 μm

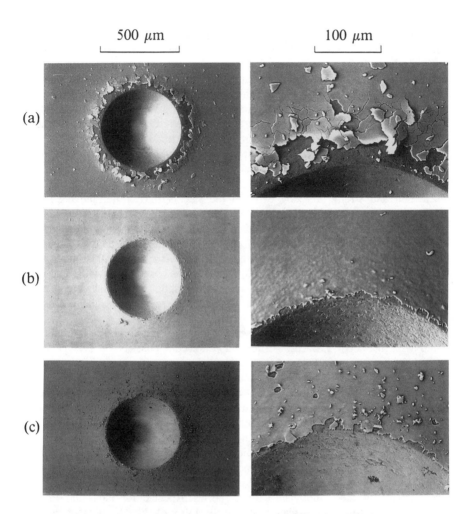

Figure 3. Two magnifications of SEM micrographs of the Rockwell indentations performed on the MoS_2 films deposited on differently treated 440C. The treatments were (a) solvent cleaning, (b) acid etching, and (c) base washing.

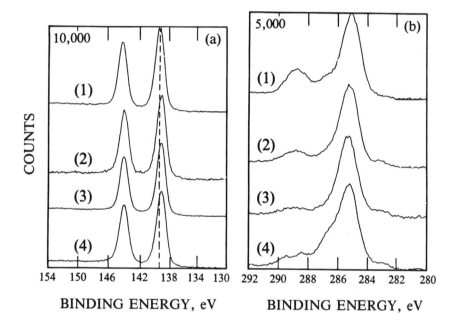

Figure 4. The (a) Pb $4f_{7/2, 5/2}$ and (b) C 1s XPS peaks of several Pbnp treated 440C samples. The samples are: (1) solvent cleaned, unscratched; (2) solvent cleaned, scratched; (3) acid etched, unscratched; and (4) based washed, unscratched.

Pbnp through the breaking of the C-O bonds in the carboxylate group. The resulting Pb binding energy is not consistent with any forms of lead oxide and, coupled with the remaining carboxylate intensity, indicates that the chromium oxide species exposed by the chemical treatments and the metals exposed to the Pbnp solution during the scratching process are chemisorbing part of the Pbnp molecule.

Figure 5 shows the Pb 4f peaks from several Pbnp treated steel surfaces as a function of temperature. The solvent cleaned sample (a) loses most of its Pb signal on heating in UHV, indicating desorption of the Pbnp. The remaining Pb intensity shifts to two lower binding energy species indicating some chemisorbed Pbnp (138.8 eV) and the formation of metallic lead (136.6 eV). In contrast, the scratched region of this surface (b) retained most of its Pb as metallic lead. Both the acid etched (c) and base washed (d) surfaces retained some metallic lead and chemisorbed Pbnp after heating. The reaction to metallic Pb, however, occurs at a lower temperature on the acid etched surface and this surface also retains more total Pb than the base treated surface. In fact, other alkaline washed samples that had lower Fe/Cr ratios than the sample from which spectra are shown in Figure 5 retained no measurable amount of Pb. The steel surface composition also changes with heating, with metallic iron emerging with increasing temperature. The reaction of Pbnp to metallic Pb is strongly correlated to the emergence of metallic Fe.(17) This suggests that Pbnp reacts with metallic Fe under EP conditions to provide a layer of metallic lead that acts as a solid boundary lubricant.

Conclusions

The surface chemical composition of 440C steel has been explored and modified by chemical and physical treatments. The layered oxide structure for a sample polished in air with an alumina/water slurry and cleaned with ethanol had an iron oxide layer ~ 2.5 nm thick and a chromium oxide underlayer ~ 1.5 nm thick on top of the metallic steel substrate. Beneath the Cr_2O_3 layer, the steel had essentially the bulk composition, indicating that the chromium enrichment observed in other studies of high chromium steels is limited to a very narrow region in surfaces treated this way. The chemical composition of 440C can be altered by both physical and chemical means. When the surface is bombarded with Ar^+ ions, the iron oxide is both sputtered away and chemically reduced to metallic Fe. Etching the surface with HCl removes the iron oxide overlayer but can also cause surface damage by breaching the chromium oxide barrier layer and corroding the bulk steel. The alkaline wash treatment is the most effective at selectively removing the iron oxide overlayer without damaging the surface.

The interactions of sputter deposited MoS_2 films with the steel surfaces depended on surface pretreatments. The MoS_2 films had greater adhesion on the chemically pretreated surfaces indicating stronger bonding between the film and the altered substrate. In addition, the MoS_2 films were slightly denser, showing a small enhancement of the nucleation process during film growth. Both improved adhesion and densification of films have been shown to increase wear life in rolling contact testing (e.g., ball bearings). It is not clear, however, whether the film

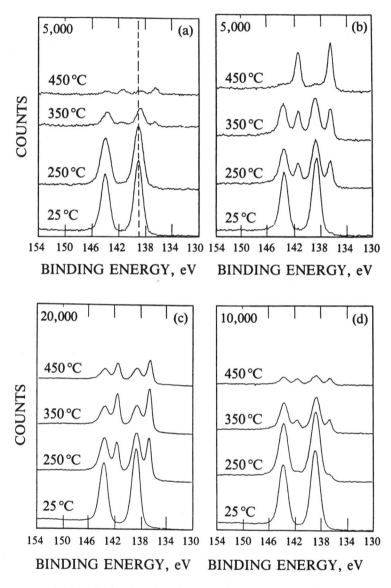

Figure 5. The Pb 4f XPS peaks of Pbnp treated 440C as a function of the sample heating temperatures listed on the figure. The samples are: (a) solvent cleaned, unscratched; (b) solvent cleaned, scratched; (c) acid etched, unscratched; and (d) base washed, unscratched.

properties were enhanced by interacting with the Cr_2O_3 underlayer or whether the chemical treatments provided more active sites for film growth.

The effects of surface composition on Pbnp bonding and reactivity are more clearly outlined by this study. Without any pretreatment, the Pbnp physisorbed on the contaminated iron oxide overlayer. On heating, most of the Pbnp present on this surface desorbed, leaving only a small amounts of chemisorbed Pbnp and metallic Pb. When the 440C surface was scratched to expose bulk steel to the Pbnp, the additive chemisorbed by breaking C-O bonds in the carboxylate group. On heating, the scratched surface readily formed metallic Pb when metallic Fe was exposed. On both the acid and base treated surfaces, Pbnp chemisorbed in a similar fashion as on the scratched area. On heating, however, the acid etched surface tended to retain more Pb and form metallic Fe and Pb more readily. In contrast, the base treated surfaces, which had the most effective removal of iron oxide, tended to lose most or all of the surface Pb. This indicates that the bonding and reactivity of Pbnp on the Cr_2O_3 underlayer is quite weak and that the surface damage done in the acid etching probably created sites for Pbnp reaction. This work suggests that Pbnp provides EP protection by forming a layer of metallic Pb when the oxide layers on asperities are worn away and temperatures increase during boundary contact. Metallic Pb can act as a solid lubricant to provide the protection.

Acknowledgments

The authors thank R. Bauer and J. Childs of The Aerospace Corporation for experimental assistance. This work was supported by the Air Force Systems Command, Space Systems Division, under contract F04701-88-C-0089.

Literature Cited

1. Jin, S. and Arens, A. *Appl. Phys. A.* **1990**, *50*, 287, and references therein.
2. Castle, J. E.; Ke, R.; and Watts, J. F. *Corros. Sci.* **1990**, *30*, 771.
3. Winer, W. O. *Wear* **1967**, *10*, 422.
4. Gardos, M. N. In *Proceedings of the 16th Leeds-Lyon Symposium on Tribology*, Dowson, D.; Taylor, C. M.; and Godet, M., eds.; Elsevier: Amsterdam, 1990, p. 2.
5. Hilton, M. R. and Fleischauer, P. D. *J. Mater. Res.* **1990**, *5*, 406, and references therein.
6. Fleischauer, P. D. and Bauer, R. *Tribol. Trans.* **1988**, *31*, 239.
7. Hilton, M. R. and Fleischauer, P. D. *Mat. Res. Soc. Symp. Proc.* **1989**, *140*, 227.
8. Didziulis, S. V. and Fleischauer, P. D. *Langmuir*, **1990**, *6*, 621.
9. Hilton, M. R.; Bauer, R.; Didziulis, S. V.; and Fleischauer, P. D. *Thin Solid Films*, **1991**, *202*, in press.
10. Mathieu, H. J. and Landolt, D. *Corros. Sci.* **1986**, *26*, 547.
11. Ertl. G. and Kuppers, J. *Low Energy Electrons and Surface Chemistry, 2nd ed.*; Verlagsgesellschaft: Weinheim, FRG, 1985, ch. 3.

The DLC and SiN$_x$ films used in this investigation were grown by means of plasma chemical depositions (9-12). The properties of thin films are sensitive to the plasma deposition conditions (13-17). Films of the most diverse composition and structure may be formed. Changes in film compositions and film properties can be perceived with the aid of analytical techniques.

The objective of this paper is to review the changes in chemistry, compositions, and properties of plasma-deposited DLC and plasma-deposited SiN$_x$ films by the plasma deposition parameter settings and the subsequent effects on the adhesion and friction properties. Analytical methods such as Auger electron spectroscopy (AES), x-ray photoelectron spectroscopy (XPS), ellipsometry, and nuclear reaction analyses are used to identify the chemical and compositional characteristics of the thin films. Sliding friction experiments were conducted to examine the friction properties of the DLC and SiN$_x$ films in contact with the hemispherical silicon nitride (Si$_3$N$_4$) pins (1.6-mm radius) in dry nitrogen and/or in a 3x10^{-8} Pa ultrahigh vacuum.

ADHESION AND FRICTION OF DIAMONDLIKE CARBON FILMS

DLC films (approximately 60 nm thick) were formed on hot-pressed, polycrystalline, magnesia-doped silicon nitride (Si$_3$N$_4$) flat substrates by using the 30-kHz ac glow discharge from a planar plasma reactor at deposition powers of 50 to 300 W (0.8 to 5 kW/m^2 (11-17). The gas source was methane (99.97 percent pure).

Two sets of sliding friction experiments were conducted using hot-pressed, polycrystalline, magnesia-doped Si$_3$N$_4$ pin specimens. In the first set, multipass sliding friction experiments were conducted with hemispherical Si$_3$N$_4$ pins (1.6-mm radius) in a dry nitrogen environment with a load of 1 N (Hertzian contact pressure, 910 MPa) and at a sliding velocity of 8 mm/min at room temperature (18). The pin was made to traverse the surface of the DLC films. The motion was reciprocal. Reference experiments for friction were also conducted with a single-crystal (111) diamond flat and uncoated Si$_3$N$_4$ flats in contact with hemispherical Si$_3$N$_4$ pins in dry nitrogen.

In the second set, single-pass sliding friction experiments were conducted with the as-coated DLC films deposited on Si$_3$N$_4$ flats in contact with ion-sputter-cleaned hemispherical Si$_3$N$_4$ pins (1.6-mm radius) in ultrahigh vacuum with loads up to 1.7 N (average Hertzian contact pressure, 1.5 GPa) and at a sliding velocity of 3 mm/min at temperatures to 700 °C (19).

Dry Nitrogen Environment

The coefficients of friction for a DLC film deposited at 50 W and for an uncoated Si$_3$N$_4$ flat are plotted as a function of the number of repeated passes (Figure 1). The friction data presented in Figure 1 indicate that the coefficient of friction was considerably lower for the plasma-deposited DLC film than for the uncoated Si$_3$N$_4$ flat. The presence of DLC films decreased friction.

Figures 2(a) and (b) present typical plots of the coefficients of friction for plasma-deposited DLC films at low (50 W) and high (250 W) deposition power as a function of the number of repeated passes. The friction characteristics of the DLC films made by different deposition powers were of two types. With the DLC films deposited at 50 to 150 W, the first type of friction characteristics (Figure 2 (a)) was generally

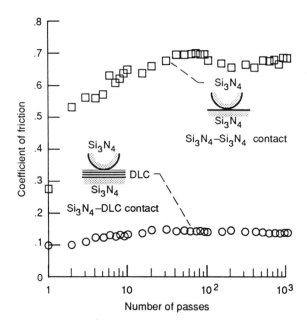

Figure 1. Comparison of coefficients of friction for 50-W plasma-deposited DLC film and for uncoated Si_3N_4 in contact with Si_3N_4 pins in dry nitrogen.

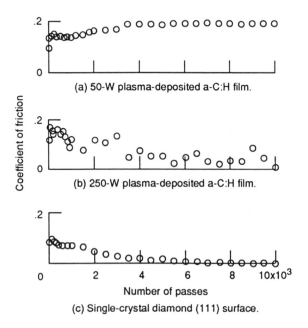

Figure 2. Average coefficient of friction as function of number of passes of Si_3N_4 pin in contact with 50-W plasma-deposited DLC film, 250-W plasma-deposited DLC film, and single-crystal diamond (111) surface in dry nitrogen.

observed. The coefficient of friction was found to increase with increasing numbers of passes. This increase, however, was small in a dry nitrogen environment even after 10 000 passes. The second type (Figure 2 (b)) was observed primarily with the DLC films deposited at 200 to 300 W. Although the coefficient of friction increased with increasing numbers of passes for about 10 passes in the dry nitrogen environment, it generally decreased in the range of 10 to 10 000 passes. At 1000 passes and above, the coefficient of friction decreased drastically. In this range the coefficients of friction for the film deposited at 250 W were much lower than those for the film deposited at 50 W.

Figure 2(c) presents typical plots of the coefficient of friction for the bulk diamond flat as a function of the number of repeated passes in dry nitrogen. In Figure 2(c), the initial friction of the bulk diamond flat in contact with the Si_3N_4 pin was low (0.08). Although the coefficient of friction increased slightly with increasing numbers of passes for about 100 passes, it generally decreased between 100 and 10 000 passes. At 1000 passes and above the coefficient of friction for the bulk diamond became very low. Comparing Figure 2(c) with Figures 2(a) and (b) shows that the friction behavior of the bulk diamond was similar to that of the DLC film deposited at 250 W. The coefficients of friction of both the DLC film deposited at 250 W and the bulk diamond in contact with Si_3N_4 pins were generally low. In dry nitrogen sliding action altered the friction behavior of both the DLC film and the bulk diamond similarly.

As shown in Figures 2(a) and (b), the friction properties of plasma-deposited DLC films can be controlled by the deposition parameter settings. To understand the changes in frictional properties by the plasma deposition conditions, some physical characteristics of the DLC films were examined (*15,21*). Auger electron spectroscopy (AES) and x-ray photoelectron spectroscopy (XPS) measurements indicated that the bulk of the DLC films contained only carbon; no other element was observed to the detection limits (0.1 at. %) of the instruments. On the other hand, the secondary ion mass spectroscopy (SIMS) depth-profiling studies indicated that the CH_x^+ (x = 0, 1, 2, 3, . . .) distributions were uniform in the bulk of the film but that oxygen was present throughout the film. This determination could not be made with the less-sensitive AES technique. Relative counts of hydrocarbon ions determined by means of SIMS depth-profiling studies also indicated that the predominant ion was CH^+; it was interesting that a higher CH^+ level was obtained from films produced at the higher power densities. Additional ions were CH_2^+, CH_3^+, C_2H^+, $C_2H_2^+$, $C_2H_3^+$ (*15,21*).

Figures 3(a) and (b) present the nuclear reaction analyses data on hydrogen concentration and the argon ion etching rate of the DLC films (*15,21*). The hydrogen concentration in the plasma-deposited DLC films is highly variable, but it generally decreases when the deposition power increases from 25 to 300 W. In Figure 3(b), the argon ion etching rate of the DLC films drops from 80 to 50 nm/min when the deposition power is increased from 25 to 300 W. Therefore, the higher the plasma deposition power, the greater the DLC film density.

The Vickers microhardness data for the DLC films deposited on Si_3N_4 substrates at various powers were obtained (*21*). All Vickers indentations were made at a load of 0.25 N at room temperature. The time in contact was 20 sec. The indentations were viewed through an optical microscope, and the sizes of the indentations were measured with a micrometer. As one might expect, the microhardness of the DLC films on Si_3N_4 substrates was greatly influenced by that of the substrates because the

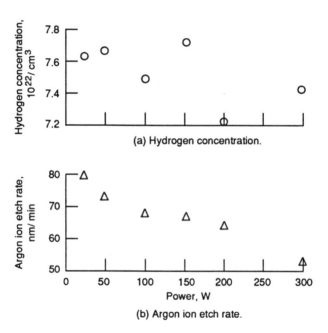

(a) Hydrogen concentration.

(b) Argon ion etch rate.

Figure 3. Hydrogen concentration and argon ion etching rate as functions of deposition power for DLC films.

load was mostly borne by the substrates. Although the DLC film thickness to indentation depth ratio was small (~0.07), the average microhardness of the DLC films increased from 2193 to 2989 (21.5 to 29.3 GPa) as the deposition power increased. Note that the average Vickers microhardness of uncoated Si_3N_4 was 1749 (17.1 GPa).

Thus, the results obtained from the physical characterization revealed that a decrease in hydrogen concentration was accompanied by an increase in film density and hardness. The friction properties shown in Figures 2(a) and (b) can be defined by hydrogen concentration. In other words, the diamondlike friction behavior (Figure 2(b)) is associated with a lower hydrogen concentration, a lower etch rate (higher film density), and a higher hardness.

To understand the drastic decrease in friction of the bulk diamond at 1000 passes and above as shown in Figure 2(c), the wear tracks on the bulk diamond surface were examined by optical microscopy and x-ray photoelectron spectroscopy (XPS). The optical photomicrograph clearly revealed that the sliding action resulted in a thin filmy product on the wear tracks and wear debris particles deposited primarily on the sides of the wear tracks. The drastic decrease in friction of the bulk diamond shown in Figure 2(c) correlated with the generation of the thin film product on the wear tracks.

The XPS spectra for C_{1s} of the as-received surface of the bulk diamond primarily included CH_x and C-C peaks as well as negligibly small C-O and C=O peaks. It is hard to distinguish C_{1s} peaks associated with C-C bondings from those associated with CH_x by XPS. A comparison of the XPS results obtained from the as-received surface of the bulk diamond and from the wear tracks formed on the bulk diamond surface indicated that the C_{1s} spectra of the as-received diamond surface was the same as those of the wear tracks. Therefore, the thin filmy product generated on the wear tracks of the bulk diamond may be a hydrocarbon-rich reaction film.

The XPS spectra for C_{1s} peaks of the as-deposited surface and of the wear tracks of the DLC film plasma-deposited at 250 W primarily included peaks related to CH_x and C-C as well as small C-O and C=O peaks. Further, the XPS spectra of the 250-W plasma-deposited DLC film were similar to those for the bulk diamond. Therefore, the decrease in friction of the DLC film shown in Figure 2(b) may also be due to the generation of a substance, probably a hydrocarbon-rich reaction film (*20*).

Ultrahigh Vacuum Environment

Typical plots of the coefficient of friction for the DLC films in contact with Si_3N_4 pins in ultrahigh vacuum are presented as a function of surface temperature in Figure 4. Comparative data for uncoated Si_3N_4 flats in contact with Si_3N_4 pins are also presented in Figure 4. With the plasma-deposited DLC films the coefficient of friction remained low at temperatures to 500 °C and rapidly increased with increasing temperatures at 600 °C and above, remaining high in the 600 and 700 °C range. When compared with the coefficient of friction for Si_3N_4 flats in contact with Si_3N_4 pins, the coefficient of friction for the DLC films in contact with the Si_3N_4 pins were generally much lower at temperatures to 500 °C. The presence of DLC films decreased friction in the temperature range. It is also interesting to note that the coefficient of friction for the DLC films had a very low coefficient of friction (e.g., about 0.08 at 500 °C for the 150-W plasma-deposited DLC film and about 0.05 at

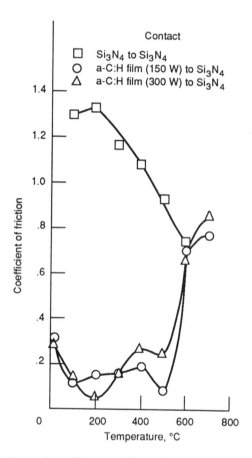

Figure 4. Comparison of coefficients of friction for DLC films and for uncoated Si₃N₄ flats in contact with Si₃N₄ pins in vacuum.

200 °C for the the 300-W plasma-deposited DLC film) even in an ultrahigh vacuum environment.

To understand the mechanism involved in the rapid increase in friction at 600 and 700 °C shown in Figure 4, the tribological properties are compared with the optical bandgap for the plasma-deposited DLC films on quartz substrates. Details of the preparation and characterization techniques and the experimental results were reported elsewhere (22) and will not be repeated here. Figure 5 presents the optical bandgap of the 150-W plasma-deposited DLC films as a function of annealing time at 400 and 600 °C. The thermal processing of the films was accomplished in a nitrogen gas with a tungsten halogen light. The main part of the reduction in the optical bandgap is obtained at short annealing time. The mechanism involved in the rapid increase in friction (Figure 4) and in the decrease in optical bandgap (Figure 5) should be related to the two-step process–namely, carbonization and polymerization (23). The carbonization stage includes loss of volatile matter, which we identified with hydrogen loss in this case (1). This stage occurs in the 400 to 600 °C temperature range in the DLC films. The polymerization stage includes the formation of graphitic crystallites or sheets. The two processes of carbonization and polymerization occur simultaneously in the DLC films.

ADHESION AND FRICTION OF SILICON NITRIDE FILMS

The material studied in this section is silicon nitride (SiN_x). Thin films containing SiN_x were deposited by high- and low-frequency plasma (13.56 MHz and 30 kHz) to a thickness of 77 nm and 86 nm, respectively. To minimize coating-substrate interactions and material variables, SiN_x was plasma-deposited on well-defined pure silicon substrates rather than on engineering alloys such as 440C stainless steel. Both adhesion and sliding friction experiments were conducted with as-deposited SiN_x films in contact with the ion-sputter-cleaned hemispherical monolithic magnesia-doped Si_3N_4 pins (1.6-mm radius) at temperatures to 700 °C in a $3x10^{-8}Pa$ ultrahigh vacuum. Details of the specimen preparation and tribological experiments are reported in reference 24.

The adhesion strength of the interfacial bonds between the surfaces of the plasma-deposited SiN_x films and monolithic Si_3N_4 pins in contact can be expressed as the force necessary to pull the surfaces apart; it is called the pull-off force. There was no significant change in the pull-off force with respect to load over the load range of 1 to 6 mN. The data, however, indicated clearly that the adhesion depended on the temperature and the plasma deposition frequency (24).

Figures 6(a) and (b) present the average pull-off forces for the SiN_x films deposited by high- and low-frequency plasmas as a function of temperature. Although the pull-off force (adhesion) for the high-frequency plasma-deposited SiN_x films increased slightly with temperatures up to 400 °C, it generally remained low at these temperatures (Figure 6(a)). The pull-off force increased significantly at 500 °C and remained high between 500 and 700 °C. Very strong adhesive bonding can take place at the contacting interface at temperatures of 500 to 700 °C.

For the SiN_x films deposited by low-frequency plasma (30 kHz), the pull-off force gradually increased with increasing temperature (Figure 6(b)). When compared with

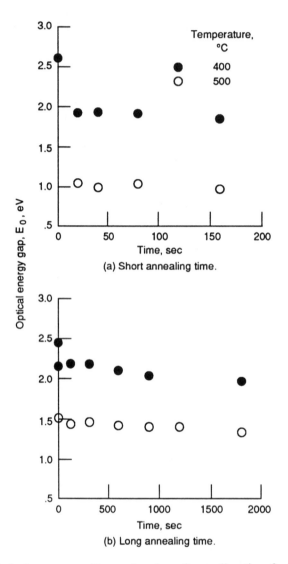

Figure 5. Optical energy gap E_0 as a function of annealing time for DLC films on quartz annealed at two temperatures.

the results of Figure 6(a), however, the pull-off forces shown in Figure 6(b) were generally lower even in the high temperature range of 500 to 700 °C.

The coefficients of static friction for the SiN_x films deposited by high- and low-frequency plasma as a function of sliding temperature are presented in Figures 7(a) and (b). The static friction characteristics are the same as those of adhesion presented in Figures 6(a) and (b). The trend for the coefficient of dynamic friction was also quite similar to that of adhesion (25). When compared with the static friction, the coefficient of dynamic friction was generally lower at temperatures to 700 °C.

To examine the effect of the plasma deposition frequency on the properties of the SiN_x films, the films were analyzed by Auger electron spectroscopy (AES) and x-ray photoelectron spectroscopy (XPS) and by ellipsometry. Comparative surface analyses were also conducted with SiN_x films deposited on gallium arsenide and indium phosphide substrates (25).

AES analyses provided complete elemental depth profiles for SiN_x films deposited by high- and low-frequency plasmas (13.56 MHz and 30 kHz) as a function of sputtering time. Typical examples are presented in Figures 8(a) and (b), respectively. Comparing the high- and low-frequency plasma-deposited SiN_x films indicated a higher silicon to nitrogen ratio for the film deposited at 13.56 MHz than for that deposited at 30 kHz.

The SiN_x films deposited by high- and low-frequency plasma were also probed by XPS. The silicon to nitrogen ratios of the argon ion-sputter-cleaned SiN_x films are presented in Table I. Since the silicon to nitrogen ratios were much greater for the films deposited at 13.56 MHz than for those deposited at 30 kHz, they supported the AES data.

TABLE I. - RATIO OF SILICON TO NITROGEN

IN SiN_x FILMS

Substrate	Plasma deposition frequency	
	30 kHz	13.56 MHz
Si	1.1	1.4
GaAs	1.2	1.4
InP	1.1	1.7

The refractive index and the absorption coefficient of the SiN_x films were also examined by ellipsometry (24). The results of the investigation indicated that the refractive index for the films deposited at 13.56 MHz was higher than that of pure amorphous Si_3N_4 film (25). This indicated that a small, but not insignificant, amount of amorphous silicon with its higher refractive index was present. In addition, pure amorphous Si_3N_4 film did not absorb at all for wavelengths above 300 nm, whereas amorphous silicon showed absorption in the 400- to 550-nm wavelength range. These facts reinforced our conclusion that amorphous silicon was present in the high-frequency (13.56 MHz) plasma-deposited SiN_x films.

The low-frequency (30 kHz) plasma-deposited SiN_x films had lower refractive indexes than did pure SiN_x and they had almost vanishing absorption. This indicated

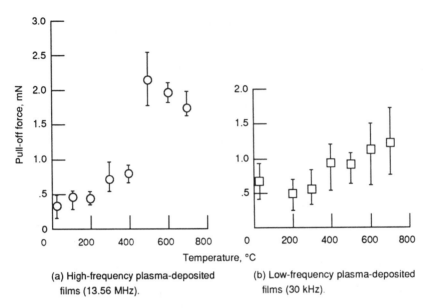

(a) High-frequency plasma-deposited films (13.56 MHz).

(b) Low-frequency plasma-deposited films (30 kHz).

Figure 6. Pull-off force (adhesion) as a function of temperature for plasma-deposited SiN_x films in contact with Si_3N_4 pins in vacuum.

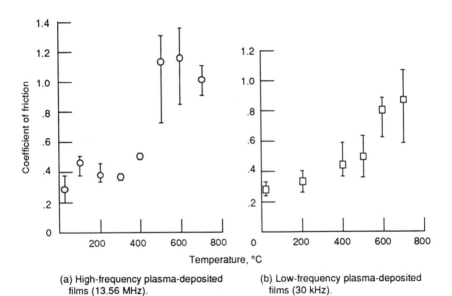

(a) High-frequency plasma-deposited films (13.56 MHz).

(b) Low-frequency plasma-deposited films (30 kHz).

Figure 7. Coefficient of friction as a function of temperature for plasma-deposited SiN_x films in contact with Si_3N_4 pins in vacuum.

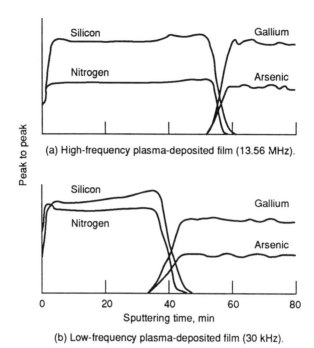

(a) High-frequency plasma-deposited film (13.56 MHz).

(b) Low-frequency plasma-deposited film (30 kHz).

Figure 8. AES depth profile for plasma-deposited SiN$_x$ films on gallium arsenide.

that the films contained either a small number of voids or a small amount of oxygen and a negligibly small amount of amorphous silicon. Note that oxygen usually exists in the form of silicon oxinitride.

Results of AES, XPS, and ellipsometric measurements of SiN$_x$ films on silicon substrates were similar to those obtained of SiN$_x$ films on indium phosphide and gallium arsenide substrates (24). Thus, the difference in the properties of high- and low-frequency deposition is due to only to the plasma, since there is no indication of substrate interaction.

Comparing Figure 8 with Figures 6 and 7 shows that the rapid increase in adhesion and friction above 500 °C can be correlated with the higher silicon to nitrogen ratio and the larger amount of amorphous silicon present on the surface of high-frequency plasma-deposited SiN$_x$ films.

CONCLUDING REMARKS

This two-part paper describes the tribological properties and physical characteristics of (1) diamondlike carbon (DLC) films and (2) silicon nitride (SiN$_x$) films. The emphasis is to relate plasma deposition conditions to the film chemistry and composition and to the adhesion and friction of the films.

First, the frictional and physical characteristics of plasma-deposited DLC films vary with deposition power. The higher the plasma deposition power, the less the hydrogen concentration and the greater the film density and the hardness. The friction behavior of DLC films deposited at 200 to 300 W is similar to that of bulk diamond in dry nitrogen. The DLC films effectively lubricate bulk Si_3N_4 surfaces at temperatures to 500 °C even in an ultrahigh vacuum environment.

Second, the tribological and physical characteristics of SiN_x films vary with deposition frequency. The silicon to nitrogen ratios for the high-frequency (13.56 MHz) plasma-deposited SiN_x films were much greater than those for the low-frequency (30 kHz) plasma-deposited SiN_x films. Amorphous silicon was present in both low- and high-frequency plasma-deposited SiN_x films. There was, however, more amorphous silicon in the high-frequency plasma-deposited films than in the low-frequency plasma-deposited films. The presence of rich amorphous silicon on the high-frequency plasma-deposited SiN_x films correlated with the increase in adhesion and friction above 500 °C in vacuum.

Finally, a physical characterization of the films by AES, XPS, ellipsometry, or nuclear reaction analyses has contributed to the understanding of adhesion and friction behavior of the films and can partially predict the tribological properties of the coatings.

ACKNOWLEDGMENT

The author wishes to thank J.J. Pouch and S.A. Alterovitz, NASA Lewis Research Center, for providing the DLC films and analytical data and V.J. Kapoor, D.M. Pantic, G.A. Johnson, University of Cincinnati, for providing the SiN_x films.

REFERENCES

1. Angus, J.C.; Koidl, P.; and Domitz, S.; in Plasma Deposited Thin Films, Mort, J. and Jansen, F., Eds., CRC Press, Boca Raton, FL, 1986, pp 89-127.
2. Aisenberg, S. and Chabot, R.; J. Appl. Phys., 1971, 42, pp 2953-2958.
3. Holland, L. and Ojha, S.M.; Thin Solid Films, 1978, 48, pp L21-L23.
4. Berg, S. and Anderson, L.P.; Thin Solid Films,1979, 58, pp 117-120.
5. Meyerson, B.and Smith, F.W.; Solid State Commun., 1980, 34, pp 531-534.
6. Moravec, T.J. and Orent, T.W.; J. Vac. Sci. Technol., 1981, 18, pp 226-228.
7. Vora, H. and Moravec, T.J.; J. Appl. Phys., 1981, 52, pp 6151-6157.
8. Banks, B.A. and Rutledge, S.K.; J. Vac. Sci. Technol., 1982, 21, pp 807-814.
9. Valco, G.J.; Biedenbender, M.D.; Johnson, G.A.; Kapoor, V.J.; and Williams, W.D.; in Symposium on Dielectric Films on Compound Semiconductors, Electrochemical Society Proceedings, 86-3, Kapoor, V.J., Connolly, D.J., Wong, Y.H., Eds., Electrochemical Society, Pennington, NJ, 1986, pp 209-219.
10. Valco, G.J. and Kapoor, V.J.; J. Vac. Sci. Technol. A, 1985, 3, pp 1020-1023.
11. Pouch, J.J.; Warner, J.D.; Liu, D.C.; and Alterovitz, S.A.; Thin Solid Films, 1988, 157, pp 97-104.
12. Warner, J.D.; Pouch, J.J.; Alterovitz, S.A.; Liu, D.C.; and Lanford, W.A.; J. Vac. Sci Technol. A., 1985, 3, 900-903.

13. Alterovitz, S.A.; Pouch, J.J.; and Warner, J.D.; in Rapid Thermal Processing of Electronic Materials, MRS Symp. Proc., Wilson, S.R.; Powell, S.R.;   and Davies, D.E.; Eds., Materials Research Society, 1987, 92, pp 311-318.
14. Pouch, J.J.; Warner, J.D.; and Liu, D.C.; NASA TM-87140, 1985.
15. Pouch, J.J.; Alterovitz, S.A.; Warner, J.D.; Liu, D.C.; and Lanford, W.A.; in Thin Films: The Relatonship of Structure to Properties, MRS Symp. Proc. 47, Aita, C.R. and Sreeharsha, K.S.; Eds., Materials Research Society, 1985, pp 201-204.
16. Alterovitz, S.A.; Warner, J.D.; Liu, D.C.; and Pouch, J.J.; in Symposium on Dielectric Films on Compund Semiconductors, Electrochemical Society Symp. Proc. 86-3, Kapoor, V.J.; Connolly, D.J.; and Wong, Y.H.; Eds., The Eelectrochemical Society, Pennington, NJ, 1986, pp 59-76.
17. Pouch, J.J.; Alterovitz, S.A.; and Warner, J.D.; in Plasma Processing, MRS Symp. Proc. 68, Coburn, J.W.; Gottscho, R.A.; and Hess, D.W.; Eds., Materials Research Society, 1986, pp 211-218.
18. Miyoshi, K. and Rengstorff, G.W.P.; Corrosion, 1989, 45, pp 266-273.
19. Miyoshi, K and Buckley, D.H.; Wear, 1986, 110, pp 295-313.
20. Field, J.E.; Ed., The Properties of Diamond, Academic Press, 1979.
21. Miyoshi, K.; Pouch, J.J.; and Alterovitz, S.A.; Mater. Sci. Forum, 1989, 52-53, pp 645-656.
22. Alterovitz, S.A.; Warner, J.D.; Liu, D.C.; and Pouch, J.J.; J. Electrochem. Soc. 1986, 133, pp 2339-2342.
23. Robertson, J.; Adv. Phys. 1986, 35, pp 317-374.
24. Miyoshi, K.; Pouch, J.J.; Alterovitz, S.A.; Pantic, D.M.; and Johnson, G.A.; Wear, 1989, 133, pp 107-123.
25. Palik, E.D.; ed., Handbook of Optical Constants of Solids, Academic Press, 1985, pp 771-774.

RECEIVED November 25, 1991

Chapter 5

Counterface Material and Ambient Atmosphere
Role in the Tribological Performance of Diamond Films

M. T. Dugger, D. E. Peebles, and L. E. Pope

Sandia National Laboratories, Albuquerque, NM 87185–5800

Diamond films are attractive as tribological coatings due to their extreme hardness, as well as thermal and chemical stability. Widespread use of these films requires a thorough understanding of how friction and wear are influenced by chemical interaction with the counterface material and the ambient atmosphere. We have evaluated the sliding contact of diamond films deposited on silicon wafers against 440C stainless steel and diamond-coated silicon carbide in pin-on-disk friction tests. The friction coefficient has been determined for these material combinations in ultrahigh vacuum, air, oxygen and water vapor atmospheres. The diamond surfaces have been characterized in situ after sliding by Auger electron spectroscopy, electron energy loss spectroscopy, and a specially developed charging probe technique to assess surface modifications that occur due to wear. The topography of the diamond surface has been characterized ex situ with atomic force microscopy. Correlations are discussed between the chemistry and morphology of the diamond surface and the tribological test conditions.

The properties of diamond make it a good candidate for application as a tribological material. Being the hardest known material, plastic deformation and hence contact area are minimized, which contribute to low friction and low wear rate. The chemical stability of diamond permits little environmental degradation, and its thermal conductivity resists large temperature increases at contact spots. Hence, thermomechanical stresses which may lead to fatigue and fracture are minimized.

NOTE: This work was performed at Sandia National Laboratories and supported by the U.S. Department of Energy under contract number DE–AC04–76DP00789.

The tribological properties of the surfaces of natural diamond have been investigated in detail for a number of years. In 1951 Bowden and Young *(1)* demonstrated the effect of adsorbed films on the friction of natural diamond. A friction coefficient for diamond on diamond below 0.1 was observed in air, while under identical load and speed conditions after degassing the specimens at high temperature in a vacuum of 10^{-4} Pa, a friction coefficient greater than 0.5 could be produced during testing in vacuum. The diamonds had a graphitic surface film due to the degassing heat treatment, which the authors believed was rapidly scraped away when sliding was begun. The authors found that after cleaning, the friction behavior could be reversibly changed between the high and low values by testing in either O_2 or dry air and vacuum. The authors proposed that the friction coefficient decrease observed when oxygen was admitted was evidence that in air a chemisorbed layer of oxygen prevented strong interfacial bonding and kept the friction coefficient low. A monotonic decrease in friction coefficient was observed as the oxygen pressure in the test vessel was increased. If the cleaned diamond surfaces were exposed to the ambient atmosphere, the friction coefficient dropped irreversibly. Bowden and Hanwell *(2)* in 1966 demonstrated that the high "clean surface" value of friction coefficient for diamond on diamond could be generated by rubbing alone. At a test pressure of 9.3×10^{-8} Pa, an initial friction coefficient of 0.1 was observed, but within several hundred passes the friction coefficient rose rapidly to 0.9 and remained constant. After a short period of inactivity, the friction coefficient would drop back to near 0.1, but subsequently returned to 0.9 to 1.0 after only 10 to 20 passes. This behavior was attributed to cleaning the surface of adsorbed contaminants by rubbing in vacuum at room temperature.

Bowden and Brooks *(3)* explored the well known dependence of friction coefficient on crystallographic direction of the diamond surface using diamond cones of different apex angle. They found that sharper cones (higher contact stress) produced a greater frictional anisotropy with sliding direction for the same load. On a (001) diamond surface, the friction coefficient in $<100>$ directions was found to be greater than that in $<110>$ directions. Frictional anisotropy was explained in terms of the distribution of shear stresses resolved on the crystallographic planes when sliding occurs in various directions. Miyoshi and Buckley *(4)* examined the contact of clean transition metals with a clean diamond (111) surface in ultrahigh vacuum, and found that the more active the metal surface (lower percent d-bonding), the higher was the adhesion between the metal and the diamond and hence the higher was the friction coefficient. In 1979 Tabor *(5)* proposed a mechanism of diamond friction to account for the experimentally observed frictional anisotropy and the dependence on slider geometry and load. In his model, an adhesion term was combined with a Coulombic term due to the surface roughness; the variation in contact stress as asperities on one surface ride over asperities on the counterface results in an energy dissipative process. The theory can account for frictional anisotropy if the asperity slopes vary with crystallographic direction. In 1981 Seal *(6)* found that for diamond sliding on itself in air, relative humidity had no effect from near zero to saturation, including flooding the sliding surfaces with water. Seal also extended Tabor's treatment of diamond asperity interaction to the case where sliding does not occur along a slope which intersects the asperity peak (i.e. a crystallographic direction).

In addition to studies of frictional anisotropy, Hayward (7) examined the effect of test ambient on the friction behavior of diamond in contact with diamond. When solvent-cleaned diamonds were rubbed together at 10^{-6} Pa, the initial friction coefficient was 0.1, but rapidly increased to 1.4. The high friction period was accompanied by extensive surface damage. In other experiments, oxygen was admitted after the high friction behavior in vacuum had begun. A decrease in friction coefficient was observed after admitting 10^{-4} Pa oxygen, but not down to the levels that were observed in air. The author attributed this to the fact that monolayer formation time at this pressure (assuming unity sticking coefficient) was on the order of the time for one pass of the stylus. An oxygen pressure giving a monolayer formation time much less than the time required for one pass of the stylus caused a drop in friction coefficient to near the value observed in air. However, the mechanism for this reduction was clouded by copious amounts of debris produced during the high friction period, which may have become trapped between the stylus and the flat specimen and reduced the contact area (and hence the friction force). When hydrogen was present at pressures such that the monolayer formation time was less than the time between passes of the stylus, the friction coefficient was also reduced. When the hydrogen pressure was lowered, the friction coefficient did not increase to values characteristic of the vacuum environment. This result is contrary to that of Bowden and Young (1), who observed no effect of hydrogen on the friction of diamond. Hayward (7) postulated that the sliding surfaces may have been exposed to atomic hydrogen due to the proximity of an ionization gauge used to measure pressure, and concluded that the hydrogen-terminated surface was durable enough to withstand excursions to lower pressure without desorption.

Earlier studies with natural diamonds occasionally produced conflicting results when the effect of test ambient on the friction coefficient was explored. Natural diamonds may have different surface properties, depending upon how they are gathered and polished. Diamonds taken from hard rock mines are frequently collected by washing the mined debris over grease tables; the diamonds are hydrophobic and preferentially stick to the grease. This technique can not be used to gather diamonds from sandy river deposits, since these diamonds tend to be hydrophyllic. The surface properties of diamond are dependent upon the surface termination (see discussion). Polishing diamond with diamond powder in a hydrocarbon oil would produce a hydrogen-terminated surface, while polishing against an iron wheel in air would yield an oxygen-terminated surface.

Samuels and Wilks (8) demonstrated that on both the (001) and (011) faces of diamond, the friction in < 100 > directions passes through a minimum as pressure is increased, and then rises. The authors proposed two mechanisms of friction whose contributions vary differently with pressure and could explain the observed behavior. This may also explain the contradictory results obtained in earlier work on the dependence of friction on contact stress. Jahanmir and coworkers (9), in wear tests of SiC on CVD (chemical vapor deposition) synthesized diamond films, observed smoothing of the diamond coating after tests in air.   The authors proposed that contact temperatures at diamond asperities may be sufficient to cause local transformation of the diamond to graphite, and result in the observed low friction.   This transformation occurs at approximately 900 K in environments containing oxygen, and above 1800 K without oxygen *(10)*.   Gardos and Ravi *(11)* used a combination of elevated temperature and ambient control to examine the friction behavior of pure and graphite-contaminated CVD diamond in contact with pure diamond-coated and bare SiC.   Pure diamond on diamond test data was difficult to obtain due to delamination of the CVD diamond coatings on Si (100), but the results suggest that at 10^{-3} Pa (turbomolecular pumped column of a scanning electron microscope) and room temperature, the friction coefficient begins at 0.5 to 0.8 and drops within several hundred cycles.   This is due, according to the authors, to the removal of adsorbed oxygen and water vapor. Blau et al. *(12)* examined the effect of CVD diamond film morphology on the friction and wear behavior in contact with 52100 steel and sapphire. Films with pyramidal (111) facets, microcrystalline "cauliflower" morphology, and plate-like films consisting of primarily (100) platelets could be produced by manipulation of the deposition conditions.   Diamond films of all types caused cutting and accumulation of transfer material. The lowest friction coefficient was obtained for surfaces composed primarily of (100) crystal faces.   The authors stated that low friction and wear surfaces require improvements in CVD diamond film growth conditions to produce smooth films.   Hayward et al. *(13)* established a correlation between friction coefficient and the final roughness of CVD diamond surfaces. They found that diamond styli of different initial surface roughness approach the same value of friction coefficient after sliding.   This was attributed to a wear-induced topography change such that the wearing surfaces approach equivalent surface roughness.   In a recent paper, Hayward *(14)* stated the need for ultrahigh vacuum friction measurement coupled with surface science studies to resolve the issue of whether surface graphitization and reconstruction occurs at asperity tips.

There has been renewed interest in the tribology of diamond with recent advances in diamond coating technology.   While tribological elements made from monolithic diamond or sintered compacts may be prohibitively expensive and may lack required structural properties, diamond coatings on other materials provide solutions to many practical problems. Chemical vapor deposition offers a method of coating surfaces with a large fraction of sp^3-bonded carbon. Such coatings may be deposited with highly faceted surfaces appropriate as cutting or grinding media, as well as in microcrystalline form with a low surface roughness which is more compatible with applications requiring a wear-mitigating coating.   Several challenges remain in diamond coating technology. A major impediment to more widespread use of diamond films is substrate deposition temperature.   Silicon, refractory metals and some ceramics are coated regularly at substrate temperatures of 500°C and above. For good quality diamond films, present deposition methods require substrate temperatures so high that steels lose their desirable mechanical properties. Furthermore, the rate of diamond deposition usually must be limited to minimize the incorporation of graphitic inclusions, which compromise the film's mechanical and optical properties.

Before CVD diamond films can gain acceptance as tribological coatings, the effect of interaction of the diamond surface with counterface materials and the atmosphere on tribological performance must be understood. In earlier work in our laboratories, the adhesion of microcrystalline CVD diamond films to silicon substrates was investigated, techniques were developed to probe the surface chemistry of diamond films without altering it, and preliminary studies of 440C sliding against diamond in air were performed (15-16). In the present work, tribological evaluation of CVD diamond has been performed in ultrahigh vacuum, oxygen, and water-vapor atmospheres, and CVD diamond coatings sliding on themselves have been investigated.

Materials and Experimental Procedures

Specimen Preparation. Pins of 440C stainless steel were cut from rod 1.58 mm in diameter and hemispherical tips of radius 0.79 mm were machined at each end. The pins were separated by glass wool, sealed in glass tubes filled with argon, and heat treated to produce a hardness of 58-62 Rockwell C. The hemispherical surfaces were metallographically polished to an arithmetic surface roughness R_a of 2.6 +/- 0.9 nm after heat treatment to remove scale and contamination resulting from the heat treatment. Stainless steel specimens were ultrasonically cleaned in a detergent solution, acetone, and isopropyl alcohol and then rinsed in deionized water.

A diamond-coated pin was prepared using an α-SiC rod 10 mm long and 2 mm in diameter, with a polished hemispherical tip of 1 mm radius of curvature. The DC plasma-assisted CVD (DC-PACVD) method with a methane+hydrogen environment was used to grow approximately 40 μm of diamond on the pin's hemispherical surface.

Two types of flat specimen, both consisting of diamond coatings deposited on Si(100) wafers, were used in the present work. Diamond triboflats designated here as "rough" were grown using the same conditions as those described for the pin specimen above. A diamond film 0.8 μm thick was deposited on the wafer surface, and then square coupons 25 mm x 25 mm were diced from the wafer. Triboflats designated as "smooth" were deposited using a similar method, producing a 2.5 μm thick coating, and then specimens 12 mm x 12 mm were diced from this wafer. All diamond films were commercially deposited (by the Crystallume Corporation, Menlo Park, California) and utilized a thin diamond-like carbon layer at the interface to enhance the nucleation and growth of diamond and improve adhesion. No subsequent cleaning of the diamond coatings was performed. All specimens were stored in a desiccator prior to testing.

Friction Tests. All friction measurements were performed on an in situ tribometer designed to replace the conventional sample carousel in a Physical Electronics model 590 Scanning Auger Microprobe (SAM) (17). Hence Auger electron spectroscopy (AES) and electron energy loss spectroscopy (EELS) were available to diagnose the surface chemistry in situ following vacuum testing. Experiments were performed at 0.24 to 0.34 N applied normal force and 25 to 60 mm/sec sliding speed (depending upon the wear track diameter on the flat) in a unidirectional, multiple-pass mode. The friction coefficient was recorded continuously during sliding. The fixed rotational speed of the test device results in a time between passes of the slider over the same region of the diamond surface of 1.1 seconds.

The test procedure consisted of calibration of the tribometer's strain gauges, followed by establishment of a steady-state friction coefficient in laboratory ambient air. The test device was then inserted into the vacuum chamber without moving the specimens in contact, and the chamber was evacuated without baking. The vacuum system was rough-pumped with a combination of mechanical and turbomolecular pumps, which were then valved off so that the chamber was ion-pumped alone. Pressures in the mid 10^{-7} Pa range were obtainable in 24 hours without baking. Alternate test atmospheres were obtained by admitting gases through ultrahigh vacuum leak valves. Research purity oxygen was connected to one leak valve. The vapor in equilibrium above purified water in a Pyrex vial connected to another leak valve was used as the water vapor source in these experiments.

Friction tests were generally continued just until the friction coefficient equilibrated in the test ambient. Sliding was then stopped, the chamber evacuated, and surface analysis performed. In cases where sliding was performed at low gas pressures, the equilibrium friction coefficient in ultrahigh vacuum was reestablished just prior to introducing the test gas. This allowed the observed friction coefficients to be attributed to interaction of the surface with the gaseous species admitted rather than to adsorption of contaminants from the vacuum system.

Chemical Analysis. Analysis of the worn and unworn surfaces in the SAM was performed under conditions which have been established to produce no electron beam-induced contamination or chemical state changes in the diamond film *(15)*. AES, EELS, and a specially developed charging probe technique were used to quantify wear track and off-track composition as well as qualitatively assess the degree of diamond bonding in the film. In all these electron beam-based techniques, pulse counting detection was used with energy analyzer resolution set at 0.6 percent.

The charging probe technique makes use of the fact that sp^3-bonded (diamond) carbon films are insulating and sp^2-bonded (graphite) carbon films are conducting. A 5 keV, 5 nA electron beam 1.6 μm in diameter was directed at the specimen surface and an Auger spectrum was acquired in the energy region of the carbon KVV peak. In the absence of specimen charging, the carbon KVV peak occurs at 273 eV. On high quality insulating diamond surfaces, the build up of surface charge causes the peak to shift to higher kinetic energy, up to 350 eV in some cases. A partially conducting surface will result if some of the surface carbon bonds are graphitic, and cause a reduction in the peak energy. The position of the carbon KVV peak using this probe was found to be very sensitive to the degree of diamond surface damage and contamination. The degree of charging (and hence the diamond character) of as-received diamond surfaces could be increased by a light sputtering using 200 eV Ar^+ ions for 20 minutes. This treatment cleans the diamond surface of atmospheric contaminants. Heavy sputtering (2500 eV for 30 minutes) resulted in sufficient surface damage to produce a carbon KVV peak identical to that for graphite. It should be noted that the charging probe technique is sensitive only to the local conductivity and not the microstructure. Therefore, structural variations (such as amorphous or glassy carbon) which may exhibit higher conductivity than diamond would be characterized as having increased graphitic character relative to diamond.

Values of relevant electron spectroscopic data obtained from polycrystalline graphite and natural diamond are presented in Table I for reference. Spectroscopic data on tribologically-tested diamond surfaces were obtained from discrete spots approximately 1.6 µm in diameter on the flat specimen, both inside and outside the wear track. Typically 3 points outside and 9 points inside the wear tracks were analyzed after each interval of the friction test corresponding to a particular environment. In this way, surface chemical changes were monitored as sliding in alternate environments was performed.

Table I. Electron Spectroscopic Data for Reference Materials

Surface	C_{KVV}, eV		$\sigma + \pi$ Loss, eV
	Position	Width	
Polycrystalline Graphite	273	22	26.9
Natural Diamond	>350	14	33.3

Source: Adapted from ref. 16.

Profilometry. Worn and unworn surfaces were examined by stylus profilometry. For non-destructive analysis of the pin specimen, a Sloan Dektak 3030 profilometer was used, having 0.1 nm vertical resolution using a diamond stylus with a 2.5 µm radius tip and 10 mg (0.1 mN) load. The geometry of the diamond-coated flats permitted them to be examined using atomic force microscopy (AFM). A Digital Instruments Nanoscope II AFM was used, employing a scan head with 75 µm horizontal and 5 µm vertical range. Due to the extremely small probe used in the AFM (a Si_3N_4 pyramid), it has much greater horizontal resolution and can image small asperities on the diamond surface more accurately than the Dektak. Topographical data was calculated from AFM images using 20 equally-spaced lines both parallel and perpendicular to the sliding direction. For each line, a linear least squares fit of the data was subtracted from the line to remove artifacts due to sample tilt during image acquisition. The arithmetic roughness R_a in each direction was calculated by determining the average deviation of the profile from the line of best fit. The average heights, slopes and curvature of the summits were also determined in each direction. Summits are defined as portions of the profile with positive height values after subtraction of the line of best fit.

Summary of Previous Work

Samples of the smooth wafer were previously evaluated using a number of techniques *(15-16)*. Transmission electron microscopy revealed a sharp diamond diffraction pattern over at least 98% of the sample surface, with occasional fine graphitic inclusions. A strong diamond peak at 1330 cm^{-1} was observed using Raman spectroscopy, as well as a broad band centered at 1580 cm^{-1} corresponding to graphite. Another broad band near 1500 cm^{-1} is attributed to amorphous carbon, according to the model of Li and Lannin *(18)*. The non-diamond bands observed on this specimen are attributed to the small graphitic inclusions observed by TEM and to the diamond-like carbon adhesion layer deposited prior to the diamond film, observed through the relatively transparent diamond layer. The intensity of the diamond vibrational mode relative to those from other forms of carbon may not be representative of the true percentage diamond character of the film, since these non-diamond structures have a large scattering cross-section for the laser light relative to diamond. The bulk hydrogen content of the film was measured by elastic recoil detection at 1 to 3 at.%, the higher content corresponding to the radial center of the wafer. The outer 0.1 μm of the film was typically 1 at.% richer in hydrogen than the bulk, and this could be removed by vacuum annealing of the film. Acoustic emission scratch tests show two failure loads, at 2.6 and 11.2 N. The lower load failure is attributed to partial film spallation between cracks, and the higher failure load to complete film spallation. These modes may be related to cohesive failure in the silicon substrate, which gave acoustic emission peaks during similar tests at 4.1 and 10.0 N. Friction tests of the smooth diamond films against 440C stainless steel in air at ambient temperature and 20% relative humidity revealed large variations in starting and final friction coefficients. A test on a specimen from the intermediate region (between the radial center and the edge) of the wafer began at a friction coefficient of 0.20 and slowly increased to 0.40 after 4000 passes. On a specimen closer to the wafer edge, the initial friction coefficient was near 0.60, then either decreased slowly to about 0.55 or increased rapidly to 0.95 and remained constant for 4000 passes. No clear correlation was observed between friction coefficient and the degree of diamond bonding in the film as measured by the σ+π plasmon peak position, C_{KVV} peak width, degree of specimen charging, or Raman diamond to graphite peak ratios. There was a correlation between friction coefficient and oxygen concentration in the wear track (increasing friction with increasing oxygen concentration), and a much stronger relationship between friction and wear track roughness (increasing friction with increasing wear track roughness). The relationship between friction coefficient and roughness of diamond surfaces has been observed previously *(7, 12-13)*. For steel sliding on diamond, large wear track roughness, high oxygen content and high friction coefficient are probably the result of wear of the steel counterface, which leads to a large contact area and oxidized wear debris and transfer material on the diamond surface. Weak trends observed between friction and the type of carbon bonding inferred from the other analytical techniques may also be attributable to the presence of oxidized debris from the steel on the diamond surface, rather than an actual change in the type of diamond bonding. Profilometry traces exhibited positive deflections of 0.2 to 1.0 μm at the sides of the wear tracks due to accumulated wear debris, but no loss of diamond was observed.

Results

Raman Spectroscopy of As-Deposited Diamond. Raman spectra for diamond and graphite reference materials are shown in Figure 1a. The 514.5 nm line of an argon laser was used as the excitation source. We observed a single peak at 1330 cm^{-1} for a natural, gem-quality diamond. Polycrystalline graphite gave two peaks, at 1353 cm^{-1} and 1580 cm^{-1}. The Raman spectra obtained from the CVD diamond coatings used in this work are shown in Figure 1b. The data indicate the presence of a diamond band at 1330 cm^{-1} in all cases, although this peak is somewhat larger relative to the non-diamond bands for the rough diamond coating on Si(100) and the diamond coating on the α-SiC pin. The fact that the rough diamond coating on Si(100) is only about one-third the thickness of the smooth diamond coating on Si(100) suggests that the thinner film is of higher quality.

Profilometry of As-Deposited Diamond. AFM images of the diamond flats are presented in Figures 2 and 3. Figures 2a and 3b are plotted on the same scale so that the features on these surfaces may be compared directly. Contour maps are projected onto the horizontal plane in each image to illustrate the shape and extent of the diamond asperities. The smooth diamond flat in Figure 2 has an arithmetic roughness R_a of 37 nm, based on the analysis of 60 μm x 60 μm images. The higher resolution image of Figure 2b indicates that the diamond asperities are made up of many smaller asperities, giving these films the "cauliflower" morphology frequently observed in scanning electron microscope images. The large asperities are on the order of 0.3 to 0.5 μm in diameter near their base. The rough diamond flat is shown in Figure 3. Analysis of a 60 μm x 60 μm area of this surface yields an arithmetic roughness of about 188 nm, and the large asperities in this case are approximately 2 μm in diameter at the base, as shown in Figure 3b.

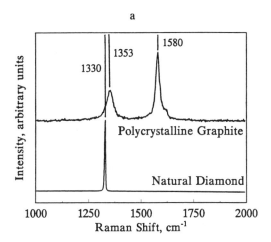

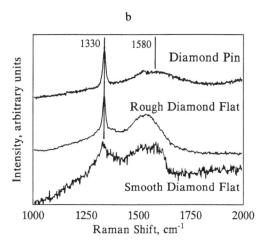

Figure 1.   Raman spectra for natural diamond and polycrystalline graphite reference materials (a), and for the diamond coatings used (b).

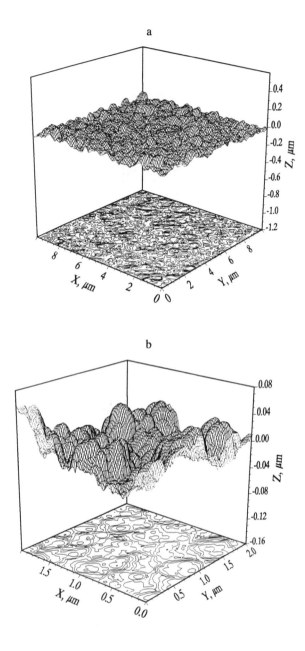

Figure 2. Atomic Force Microscope (AFM) image of a 10 μm x 10 μm area of the smooth diamond surface (a). Each of the contour lines in the horizontal projection represents a height change of 20 nm. In (b), the magnification has been increased to reveal the shape of diamond asperities. Contour lines in the horizontal projection in (b) correspond to a height change of 10 nm.

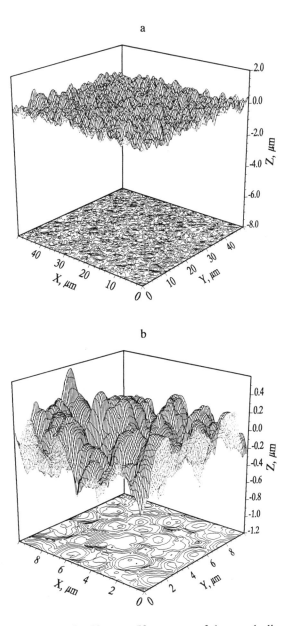

Figure 3. AFM image of a 50 μm x 50 μm area of the rough diamond surface (a). Contour lines are at 200 nm intervals. In (b), a 10 μm x 10 μm section of the surface is shown with the contour lines in the horizontal projection at 100 nm intervals.

Upon just beginning to scan the wear track regions on flat specimens, a noisy AFM image was usually obtained. The image degradation did not correspond to poor spatial resolution of the diamond crystallites, but rather to rapid and localized stylus deflections scattered over the surface. This type of image degradation was not observed on unworn regions of the surface, where a high-quality image could be obtained almost immediately after scanning was begun. On wear track regions, the horizontal scan size and scan rate were initially set at maximum, and the stylus was allowed to raster rapidly over the surface (approximately 15 ms per line, 400 lines per frame). After a time, the localized deflections cleared and a noise-free image of the surface could be obtained. This procedure was recommended by the equipment manufacturer as a method of sweeping the surface free of loose material that may cause erratic behavior of the stylus during scanning. It was found that small regions of noise within a larger image could also be removed by zooming in on the area and performing the rapid scanning as described above. This suggests that the friction tests produced wear tracks that were covered by debris which was not strongly bonded to the surface, and which could be swept aside by rapid scanning. The rapid scanning procedure was used throughout the present work to clean surfaces of loose debris so that an accurate measurement of the topography of the worn surface could be made.

The Dektak stylus profilometer was used to measure the roughness of the diamond coating on the α-SiC pin prior to tribological testing. A 2.5 μm radius stylus was used, and an R_a of 510 nm was obtained. The true roughness must be greater than this, since the tip of the large stylus is too wide to descend into narrow valleys on the surface. The coating on the α-SiC pin was therefore the roughest of all the CVD coatings examined in this study.

Stainless Steel Against Diamond-Coated Silicon. The friction coefficient versus cycle data from experiment SSD1 is shown in Figure 4. A smooth diamond flat and 440C stainless steel pin were used for this experiment. In 56 % relative humidity air the friction coefficient began at 0.24, rose rapidly to 0.38 within 20 cycles, then rose more slowly to approximately 0.45 at the end of 500 cycles in air. A large variation in the friction coefficient within each individual cycle was observed (denoted by the dotted lines in Figure 4) during this portion of the test. The test was stopped, and the tribometer was mounted on the vacuum system. The chamber was evacuated to a pressure below 7×10^{-7} Pa, and sliding was resumed. In ultrahigh vacuum, the friction coefficient on the same track began at 0.19, then rose to 0.32 after a total of 900 cycles. The fact that the minimum friction coefficient observed during a cycle was relatively unchanged in vacuum may indicate that only a portion of the wear track was damaged during the high friction period in air.

A flat was worn at the tip of the hemispherical 440C pin due to cutting by the diamond surface during this experiment. AFM imaging inside the wear track on the diamond coating revealed the surface topography shown in Figure 5. No evidence of fracture, spallation or excessive wear of the diamond surface was found, but close examination of the image in Figure 5 reveals a few very tall

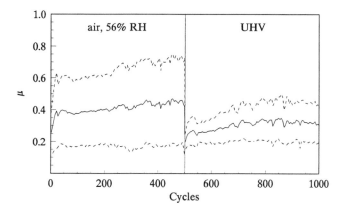

Figure 4. Friction coefficient versus number of cycles for test SSD1. 440C stainless steel sliding against smooth diamond coated silicon at 23.6 mN applied load. Solid line: mean value; dotted lines: extreme values. The vertical line indicates where the test ambient was changed.

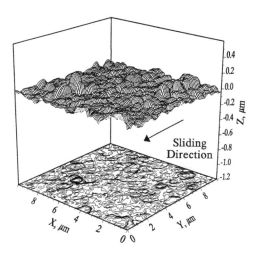

Figure 5. AFM image of a 10 μm x 10 μm area within the wear track on the diamond surface from test SSD1. Contour lines in the horizontal projection are 20 nm apart.

asperities on a surface which is otherwise identical to Figure 2a. These high areas are believed to correspond to material transferred to the diamond surface, since features of this size were not seen in any images of the unworn surface. Numerical analysis of the wear track profiles revealed that the diamond surface was smoothed slightly in spite of the new tall asperities, as shown in the Table II values for test SSD1. This smoothing is represented by a decrease in summit height, a decrease in summit slope, and a decrease in summit curvature relative to the unworn surface. The fact that the mean summit height is greater than the mean deviation from the profile average (R_a) indicates that the summits have steeper sides than the valleys. The highest asperities in Figure 5 probably have little impact on the average surface values in Table II, because the features are localized to a small portion of the surface.

Table II. Topographical Data for Smooth Diamond Film

Parameter	Unworn	SSD1[a]	SSD2[b]
R_a, nm X[c]	37.4	33.8	43.2
Y	37.7	33.7	35.9
Summit			
Height, nm X	43.2	40.4	38.3
Y	42.9	40.5	33.3
|slope|, mrad X	1.34	0.97	0.54
Y	1.23	0.93	0.50
Curvature, mm^{-1} X	-7.80	-5.28	-2.65
Y	-7.71	-4.70	-2.51

[a] Test SSD1 was performed using a 440C stainless steel counterface sliding against the smooth diamond film in air with 56% relative humidity, followed by ultrahigh vacuum.

[b] Test SSD2 was performed using a 440C stainless steel counterface sliding against the smooth diamond film in air with 55% relative humidity, followed by ultrahigh vacuum and then 1.3×10^{-4} Pa oxygen.

[c] X refers to values across the wear track (perpendicular to the sliding direction), and Y denotes the direction along the wear track (parallel to the sliding direction).

To further examine the reduction in friction observed between sliding in air and in ultrahigh vacuum on the same track, the above experiment was repeated.

The friction data from this test is shown in Figure 6. In this case the friction coefficient started at 0.34 and rose to 0.60 within 60 cycles of sliding. As observed previously, the friction continued to rise slightly with sliding in air to a value of 0.71 after 500 cycles. Surface analysis of the wear track formed in air was performed at this stage. The friction coefficient when sliding was resumed in ultrahigh vacuum was again lower than the preceding value in air, starting at 0.45. After 500 cycles in ultrahigh vacuum, the friction coefficient was 0.54. Chemical analysis of the wear track and adjacent regions was performed once more. When friction testing was resumed, an additional 70 cycles were performed in ultrahigh vacuum in order to reestablish surface conditions representative of sliding at ultrahigh vacuum prior to changing the test ambient. The friction coefficient in this stage was slightly higher than that observed at the end of the ultrahigh vacuum portion of the experiment, at 0.60. Sliding was again stopped, and pure oxygen was admitted to the vacuum system to a pressure of 1.3×10^{-4} Pa. This pressure gives a monolayer formation time (assuming unity sticking coefficient) approximately the same as the time to complete one sliding cycle. When sliding resumed after several minutes exposure to oxygen, the friction coefficient started at 0.60 and rose to 0.67 after an additional 330 cycles. An AFM image of this surface is shown in Figure 7. Topographical data from the wear track after test SSD2 indicates a substantial change from the unworn surface (Table II). The arithmetic roughness perpendicular to the sliding direction increased, while that along the sliding direction decreased slightly after the test. More dramatic changes were observed in the summit slope and curvatures. The absolute value of summit slopes decreased by a factor of two, and summit curvature decreased by approximately three times compared to the unworn surface. Changes in summit slope and curvature were relatively isotropic.

The electron spectroscopic data from this test surface is shown in Table III. Each entry is the average of 9 points if inside the wear track and 3 points if outside the wear track, except where noted. On transition from natural diamond to graphite, the charging peak (C_{KVV} position) and $\sigma + \pi$ plasmon loss energies decrease, while the width of the C_{KVV} peak increases. For ease in identifying the trends in carbon bonding type, the "percent diamond character" corresponding to the observed peak energy or width has been calculated using the data in Table I as reference data for graphite and diamond. The calculated bonding type values appear in parenthesis under the energy values in Table III. Sliding in air for 500 cycles shows a slight shift toward diamond in the charging peak position and the width of the carbon peak, but the plasmon peak shows a dramatic shift toward graphitic character. This shift cannot simply be interpreted as a change from diamond to graphite, however. The loss energy is only a function of electron density. Oxygen, being more electronegative than carbon, will tend to pull electrons away from the carbon nucleus and thus lower the average electron density near the carbon atom and consequently give a lower $\sigma + \pi$ loss energy. Therefore, high oxygen content in the wear track would cause an apparent shift of the surface binding state toward graphite as measured by ELS, while the actual environment of the surface carbon may remain sp^3. The atomic composition data

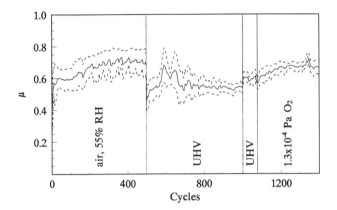

Figure 6. Friction coefficient versus number of cycles for test SSD2. 440C stainless steel sliding against smooth diamond coated silicon at 23.6 mN applied load. Solid line: mean value; dotted lines: extreme values. Vertical lines indicate where the test ambient was changed.

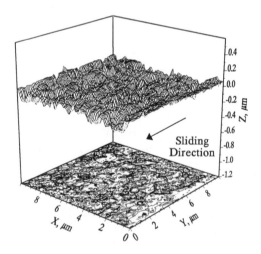

Figure 7. AFM image of a 10 μm x 10 μm area within the wear track on the diamond surface from test SSD2. Contour lines in the horizontal projection represent a height change of 20 nm.

presented in Table III indicate that this is likely the case. An equivalent number of cycles in ultrahigh vacuum on the same area of the surface produced a large reduction in the amount of sample charging, and a significant reduction in carbon peak width. The shift in the plasmon peak is consistent with the decrease in oxygen concentration observed in this sliding interval. Further cycles in ultrahigh vacuum followed by 1.3×10^{-4} Pa oxygen caused little change in the peak values from the ultrahigh vacuum test. If the plasmon peak data is disregarded due to the influence of oxygen, we have the C_{KVV} peak energy indicative of a shift toward graphite, while the width of the peak indicates a shift toward diamond. The data therefore suggest opposite trends in the carbon surface binding state with sliding.

Table III. Electron Spectroscopic Data for Test SSD2[a]

Condition	μ, final	C_{KVV}, eV Position	Width	$\sigma + \pi$ loss, eV	at.%[b] O	Fe
Unworn[c]	-	326 (69)	19 (38)	31.5 (72)	7.8	0.0
500 cycles, 55% RH air	0.71	332 (77)	18 (50)	28.0 (17)	20.0	11.4
+500 cycles, UHV	0.54	299 (34)	16 (75)	30.6 (58)	13.8	8.7
+70 cycles, UHV +330 cycles,[d] 10^{-4} Pa O_2	0.68	295 (29)	16 (75)	29.7 (44)	9.0	3.4

[a] The numbers in parenthesis indicate the percent diamond character indicated by each of the energies presented, using the data in Table I as a reference.
[b] Concentrations calculated using published relative sensitivity factors for C, O, and Fe.
[c] In all but one of the points analyzed outside the wear scar, multiple peaks were obtained after detailed scans of the C_{KVV} region, making width determination impossible.
[d] After the run in low oxygen pressure, only one location in the wear track contained iron.

The unworn surface contains only carbon and about 8% oxygen. Oxygen on the unworn surface is attributable to species adsorbed from the air during handling after deposition. After 500 cycles in air, the oxygen concentration at the surface within the wear track increases to 20%, and 11% iron is also present. The iron and most of the additional oxygen observed are the result of oxidized iron wear debris on the diamond surface. A reduction of both oxygen and iron concentration occurred after sliding for 500 cycles in ultrahigh vacuum on the same region of the diamond surface, with further reduction after an additional 70 cycles in ultrahigh vacuum followed by 330 cycles in 1.3×10^{-4} Pa oxygen. At the conclusion of the test, the oxygen concentration was close to the value observed on the unworn surface, and iron was seen in only one of the nine locations in the wear track examined.

Diamond-Coated α-SiC Against Diamond-Coated Silicon. The results of our first experiment using the rough diamond flat in contact with the diamond-coated α-SiC pin (test DD1) are shown in Figure 8. The friction coefficient equilibrated in 55% relative humidity air at 0.13 after 340 revolutions. When testing was resumed in ultrahigh vacuum, the friction coefficient climbed to 0.47 within 100 cycles, and the magnitude of the oscillations in friction coefficient also increased. When the test device was removed from the vacuum system and sliding continued in 52% relative humidity air, the friction coefficient immediately dropped to less than 0.10, and reached 0.08 within 50 cycles. Examination of AFM images from within the wear track revealed a surface morphology indistinguishable from the unworn surface shown in Figure 3. However, the calculated topographical data for test DD1 in Table IV indicates some smoothing of the rough diamond plate. The overall arithmetic roughness, summit height and slope decrease slightly relative to the unworn surface. Summit curvature increased somewhat. An explanation for the decrease in the average summit slope accompanied by an increase in curvature may be microfracture at the asperity tips, yielding no contribution to the average asperity slope but increasing the asperity curvature. The surface geometric changes are essentially isotropic. A flat wear scar was formed on the end of the hemispherical pin, and several individual traces across the wear flat (with the Dektak and a 2.5 μm stylus) gave a mean R_a of 40 nm, compared to 512 nm on the unworn pin surface. Surface analysis data for this test is presented in Table V. The probes for carbon type show no dramatic changes in bonding from the unworn surface, regardless of test ambient. Slight shifts in the C_{KVV} position are consistent with those in the σ+π loss energy, which indicate an increase in sp^3 bonding character after sliding in air, and sp^2 character after sliding in ultrahigh vacuum. In this case the oxygen content is consistently near 4 atomic percent, whether sliding was performed in air or ultrahigh vacuum, so the electron density argument can not account for the plasmon peak shift. The absence of silicon in the wear track is evidence that the diamond film did not break through on either the pin or the disk. The integrity of the diamond coating on the pin was independently verified by examining it with AES (after exposure to air), and no silicon was found.

The above experiment was repeated using a new region of the flat and alignment of the diamond-coated pin so that the flat was contacted by a new spot

Table IV. Topographical Data for Rough Diamond Film

Parameter		Unworn	DD1[a]	DD2[b]	DD4[c]
R_a, nm	X[d]	186.8	170.5	260.5	158.2
	Y	188.4	166.2	156.1	162.4
Summit					
Height, nm	X	188.3	171.1	257.2	156.0
	Y	189.3	165.2	167.4	164.2
|slope|, mrad	X	1.45	1.35	1.21	1.04
	Y	1.28	1.20	1.55	1.32
Curvature, mm^{-1}	X	-2.59	-3.21	-1.34	-2.14
	Y	-2.80	-3.31	-3.70	-2.89

[a] Test DD1 was performed using the diamond-coated α-SiC pin against the rough diamond film in air with 55% relative humidity, followed by ultrahigh vacuum and then air with 52% relative humidity.

[b] Test DD2 was performed using the diamond-coated α-SiC pin in contact with the rough diamond film in air with 51% relative humidity, followed by ultrahigh vacuum, 1.3×10^{-4} Pa oxygen, ultrahigh vacuum, 1.3×10^{-3} Pa oxygen, and finally air with 52% relative humidity.

[c] Test DD4 was performed using the diamond-coated α-SiC pin in contact with the rough diamond film in air with 13% relative humidity, ultrahigh vacuum, 1.3×10^{-4} Pa water vapor, 1.3×10^{-3} Pa water vapor, 1.3×10^{-2} Pa water vapor, and then air with 18% relative humidity.

[d] X refers to values across the wear track (perpendicular to the sliding direction), and Y denotes the direction along the wear track (parallel to the sliding direction).

on the hemispherical pin surface. In test DD2, the friction increase obtained upon continuation in ultrahigh vacuum of a test that was started in air was verified, as illustrated in Figure 9. In this case, a steady-state friction coefficient of 0.04 was obtained in 51% relative humidity air after 100 cycles. Restarting sliding in ultrahigh vacuum without disturbing the specimens gave an initial coefficient of friction of 0.30, which increased to 0.43 in 130 cycles. After the specimen adsorbed gases from the vacuum system for several days, the friction coefficient returned to approximately 0.30 when sliding was restarted, and then rose to 0.46 after an additional 155 cycles in ultrahigh vacuum. Auger analysis revealed the presence of silicon in circumferential scratches within the wear track after this stage of sliding (100 cycles air + 285 cycles UHV), but a large fraction of the

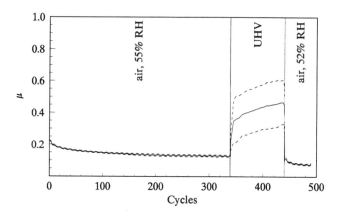

Figure 8. Friction coefficient versus number of cycles for test DD1. CVD Diamond-coated α-SiC sliding against the rough diamond-coated silicon wafer at 34.3 mN applied load in air, UHV and air. Solid line: mean value; dotted lines: extreme values. Vertical lines indicate where the test ambient was changed.

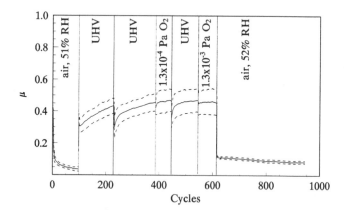

Figure 9. Friction coefficient versus number of cycles for test DD2. CVD Diamond-coated α-SiC sliding against the rough diamond-coated silicon wafer at 34.3 mN applied load in air, UHV, pure oxygen at low pressure and air. Solid line: mean value; dotted lines: extreme values. Vertical lines indicate where the test ambient was changed.

Table V. Electron Spectroscopic Data for Test DD1[a]

Condition	μ, final	C_{KVV}, eV Position	Width	σ+π loss, eV	at. % O[b]
Unworn[c]	-	296 (30)	19 (38)	29.2 (36)	4.2
340 cycles, 55% RH air	0.13	299 (34)	19 (38)	29.8 (45)	4.1
+100 cycles, UHV	0.47	293 (26)	19 (38)	28.1 (19)	4.2
+50 cycles, 52% RH air	0.08	-	-	-	-

[a] The numbers in parenthesis indicate the percent diamond character indicated by each of the energies presented, using the data in Table I as a reference.
[b] Concentrations calculated using published relative sensitivity factors for C, O, and Fe.

apparent contact path remained covered with the diamond film. An atmosphere of 1.3×10^{-4} Pa pure oxygen was established in the vacuum system and sliding resumed, with essentially no change from the vacuum friction coefficient (average m was 0.47 after 60 cycles). At this pressure, the monolayer adsorption time is about 1 second. Surface analysis of the resulting wear track again showed silicon in a narrow, circumferential scratch, but extensive delamination of the film did not occur. Another 100 cycles were run in vacuum (μ=0.47) followed by 70 cycles at 1.3×10^{-3} Pa oxygen. The mean friction coefficient at this oxygen pressure was slightly reduced from the vacuum value to 0.45. Electron microscopy after the last increment of sliding revealed a significant amount of wear debris to the sides of the wear track, and AES indicated a high concentration of silicon within the track. Hence the steady state friction from this portion of the test represents diamond sliding on silicon. When the device was removed from the vacuum system, additional cycles were run to obtain the friction coefficient for diamond on silicon in air. A value of 0.09 was obtained after 330 cycles. Surface analysis of the wear scar on the diamond-coated pin (after exposure to air) revealed no silicon, indicating that the film on the pin was still intact and that there was no transfer of silicon from the wafer to the pin. An AFM image from within the wear track is shown in Figure 10. The image has been rotated 180° relative to the other images in this paper to improve the visibility of the wear scar along the right side of the image. The deep scratch, penetrating down to the silicon substrate, is more easily visible in the contour map. This alteration of the surface morphology due to wear is reflected in the topographical data for test DD2 in Table IV. Wear along the sliding direction increases R_a and the summit heights perpendicular to the sliding direction (X), and decreases these values parallel to the sliding direction (Y), relative to the unworn surface. The summit slope and curvature data are somewhat artificial in this case, due to the localized and directional nature of the diamond film wear. A deep groove over a significant portion of the surface will cause the line of best fit to occur below the mean line of the unworn surface, thereby including regions that are actually valleys in the calculation of summit parameters. This is particularly true in the direction perpendicular to sliding. As before, the wear flat on the pin was profiled with the Dektak, resulting in an R_a of 10 nm, indicative of a high degree of polish after this test (945 cycles total) compared with the unworn pin surface.

The results of test DD4 are shown in Figure 11. In this run, the friction coefficient began at 0.32 in 13% relative humidity air. After 500 cycles in this environment, the friction coefficient decreased to 0.03. Surface analysis of the wear track was performed, and then sliding resumed in ultrahigh vacuum. The friction coefficient increased within 100 cycles to 0.33. The surface was analyzed once more, and then water vapor was introduced into the chamber. With each decade increase in water vapor pressure, the friction coefficient decreased. At 1.3×10^{-2} Pa oxygen, the friction coefficient reached a steady-state value of 0.16 for this oxygen pressure (Figure 11). Surface analysis was performed again after the last water vapor exposure. When the device was returned to 18% relative humidity air, the friction coefficient decreased further to 0.07. This sliding sequence resulted in the greatest reduction in R_a and summit height for the rough diamond surface, shown in Table IV. The electron spectroscopic data for this run is presented in Table VI. The C_{KVV} peak width is constant throughout all environments at a value close to that for graphite, and the $\sigma+\pi$ plasmon loss energy is approximately half way between that for diamond and that for graphite in all environments. The surface oxygen concentration increases slightly after the water vapor exposure, and may be indicative of adsorption of water by the diamond surface during sliding or after sliding stopped. The charging probe in these experiments exhibits similar behavior to that observed in the test with the stainless steel counterface (SSD2, Table III). Sliding in environments containing oxygen or water vapor results in an increase in charging, while sliding in vacuum seems to increase the conductivity of the diamond surface.

Discussion

440C on CVD Diamond. The rapid increase in friction coefficient observed in the first few passes of the stainless steel rider over the diamond surface in air suggest transfer of material to the diamond surface. Material transfer will increase the adhesive component of friction between the mating surfaces, and non-uniform transfer would be expected to increase the roughness-induced contribution to the friction coefficient as well. Transfer over only a portion of the wear track may

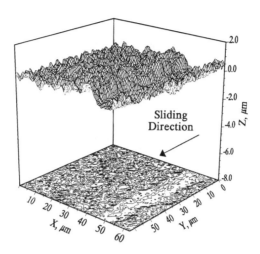

Figure 10.   AFM image of a 63 μm x 63 μm area within the wear track on the diamond surface from test DD2, showing a wear scar down the right side of the image.   Contour lines in the horizontal projection represent a height change of 200 nm.

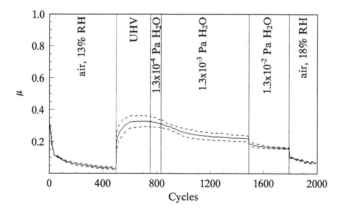

Figure 11.   Friction coefficient versus number of cycles for test DD4.   CVD Diamond-coated α-SiC sliding against the rough diamond-coated silicon wafer at 34.3 mN applied load in air, UHV, and water vapor.   Solid line: mean value; dotted lines: extreme values.   Vertical lines indicate where the test ambient was changed.

Table VI. Electron Spectroscopic Data for Test DD4[a]

Condition	μ, final	C_{KVV}, eV Position	Width	$\sigma+\pi$ loss, eV	at. % O[b]
Unworn	-	319 (60)	21 (12)	28.6 (27)	3.0
500 cycles, 13% RH air	0.03	330 (74)	21 (12)	29.4 (39)	2.9
+250 cycles, UHV	0.33	308 (45)	21 (12)	28.9 (31)	4.3
+81 cycles, 1.3×10^{-4} Pa H_2O	0.32				
+658 cycles, 1.3×10^{-3} Pa H_2O	0.23				
+303 cycles, 1.3×10^{-2} Pa H_2O	0.16	326 (69)	21 (12)	29.1 (34)	5.0
+200 cycles, 18% RH air	0.07	-	-	-	-

[a] The numbers in parenthesis indicate the percent diamond character indicated by each of the energies presented, using the data in Table I as a reference.

[b] Concentrations calculated using published relative sensitivity factors for C, O, and Fe.

also explain the constant minimum friction coefficient observed in test SSD1 (Figure 4). That transfer of steel to the diamond surface occurs in air is demonstrated by the atomic composition data in Table III.

Tests SSD1 and SSD2 indicate that large variations in the friction coefficient for tests on the same surface under identical conditions are possible (in agreement with our earlier work). However, both tests indicate that the friction coefficient for stainless steel sliding on diamond decreases when humid air is removed from the system. Two mechanisms contribute to this behavior. First, since diamond is much harder than the counterface in this case, the principal means for debris generation will be abrasive wear of the steel surface. In air, the diamond asperities will plow through a surface oxide on the steel which is much more resistant to deformation than the underlying metal. Hence greater energy must be dissipated to move the asperities through the oxide surface, resulting in a high friction coefficient. With an ample supply of oxygen and water vapor, the steel surface is continuously oxidized so that diamond asperities are predominantly sliding against iron oxide. Transfer of steel to the diamond surface also occurs, according to the composition data in Table III. Consequently, sliding of (oxidized) steel on steel also must contribute an adhesive component to the friction force. In vacuum, there is insufficient oxygen available to maintain the surface oxide on the pin, so diamond asperities plow through the softer steel surface, resulting in a lower friction force. The data in Table III indicates that sliding of a steel pin on a diamond surface in vacuum not only results in a lower oxygen concentration in the wear track than when sliding occurs in air, but iron-rich debris deposited during sliding in air appears to be removed when sliding continues over the same surfaces in vacuum. This may be due to a greater affinity of the iron oxide for the freshly exposed metal surface than for the diamond. The data in Table III also indicates that the amount of oxidized debris on the diamond surface continues to decrease after sliding in oxygen at 1.3×10^{-4} Pa. This suggests that greater oxygen pressure may be required to maintain the surface oxide on the steel. Second, a chemical change at the contact spots may also contribute to lower friction. One method of polishing diamond is to rub it against a rotating cast iron wheel or "scaife." It has been argued *(19)* that conversion of diamond to graphite at the contact spots, catalysed by iron, is a more probable mechanism for diamond wear than direct diffusion of carbon from the diamond surface into the iron. Reduced friction in vacuum at iron/diamond contact spots may also then result from local iron-catalysed transformation of diamond to graphite. Although the charging probe and ELS data support this hypothesis, the width of the C_{KVV} peak does not. Furthermore, the shifts observed in the charging probe and ELS data may be explained by transfer of steel to the diamond surface, without invoking a change in surface bonding type.

Analysis of AFM images acquired within wear tracks indicates that the summit height, slope and curvature of asperities on the diamond surface are reduced when sliding against a steel counterface. The degree of smoothing increases with the number of cycles. Since the topography of the diamond surface could not be determined without disturbing the specimens in contact, it is not known which environment promotes the greatest change in the diamond surface topography. The changes in topographical parameters of the surface may be explained in terms of wear of the diamond asperities or accumulation of debris from the steel counterface in the regions between diamond asperities. However, our surface analysis data showed iron in only one of nine locations analyzed on track SSD2 at the conclusion of sliding. If enough iron-rich debris were collected on the diamond surface to alter the surface topographical parameters, as shown in Table II, one would expect to see much more iron in the wear track. This implies that a portion of the surface topography change must be due to wear of the diamond, perhaps after first being converted to graphite as suggested by Wilks and Wilks *(19)*.

CVD Diamond on CVD Diamond. The initial decrease in friction coefficient as one microcrystalline CVD diamond surface slides over another is due to a reduction in the roughness of the coating as repeated passes are made over the same region of the surface *(13)*. Most of this change occurs in the first several cycles of sliding as an equilibrium composite surface roughness is approached. The starting friction coefficient and the number of passes required to reach a steady-state value are a function of the initial surface topography, surface chemical state, and presence of adsorbed material (see Figures 8, 9, and 10). In the present investigations, steady-state values were obtained in 100-300 passes.

The increase in friction coefficient observed when CVD diamond slides against itself in vacuum is due to the removal of adsorbed surface contaminants. This film may be removed by repeated sliding over the same track *(2)*, ion bombardment *(4)*, or heating *(11)*. Despite the fact that the sticking coefficient for molecular oxygen on natural diamond is estimated to be less than 10^{-6} *(20)*, it is believed that the surface of CVD diamond is easily oxidized in air, due to the presence of dangling bonds *(11)*. It should be noted that the ability of the diamond surface to adsorb water vapor is determined by the termination species present. Fully oxygenated surfaces are hydrophilic, whereas outgassed or hydrogen-terminated surfaces are hydrophobic *(20)*. The adhesion of cleaned diamond surfaces then arises from removal of the surface termination, followed by carbon-carbon bonding across the interface. A recent molecular dynamics model suggests that diamond surfaces may also interact adhesively if one or both are only partially hydrogen-terminated *(21)*.

The adhesive component of diamond friction has been viewed skeptically by some authors. Samuels and Wilks *(8)* see no need to invoke an adhesive mechanism for diamond friction in air since their ratchet and elastic loss mechanism can account for the observed pressure dependence of diamond friction. This conclusion is based on the well-known experimental observation that lubricating oils have little effect on the friction coefficient of diamond surfaces. However, all this really means is that either the adsorbates from air are as effective as the lubricant in reducing the friction coefficient, or the same surface adsorbed layer occurs whether the surface is in air or covered by oil. Recent work suggesting that water is effective in reducing diamond friction implies that there is indeed an adhesive component to the friction coefficient *(14)*.

There exists some uncertainty in the role of wear debris in the reduction of friction coefficient of diamond when admitting gases to a vacuum system *(14)*. The high friction behavior which occurs when diamonds are slid in vacuum has been observed to produce significant amounts of wear debris *(7)*. When gases were subsequently bled into the vacuum system, it was not known whether the resulting decrease in friction was due to reformation of surface adsorbates to reduce carbon bonding across the interface, or debris becoming trapped between the sliding surfaces to reduce the contact area and the friction force. There has even been the suggestion that hydrocarbon-like lubricating debris may form in some environments *(14)*. Figure 9 clearly shows that after idle periods when surface analysis was performed the friction coefficient between CVD diamond surfaces was reduced, but after sliding continued in vacuum the high friction behavior could be restored within a few passes of the slider. It is difficult to imagine how diamond debris or lubricating hydrocarbon debris could be ejected from the contact zone with sufficient speed and reproducibility to produce the friction data shown in Figure 9. We favor the desorption and reformation of protective surface layers to explain the observed phenomena.

Local graphitization of the diamond surface at temperatures greater than 900 K in the presence of oxygen has been proposed to account for the low friction of diamond in air *(9)*. Our electron spectroscopy data from tribological tests of diamond in contact with diamond suggest an increase in the diamond bonding character of the surface after sliding in air, accompanied by a low friction coefficient. Furthermore, a shift to graphitic character is consistently measured after sliding in vacuum where a relatively high friction coefficient is observed. These results do not support the formation of graphite at contact spots as a lubrication mechanism, and rule out a simple thermal transformation of diamond to graphite at asperity tips since this would be more difficult in vacuum. The increase in diamond character after sliding in air is due to the removal by wear of adsorbed contaminants and perhaps also a thin damaged zone produced during film deposition. The graphitic nature of the surface after wear over the same location in vacuum may be induced by the removal of surface adsorbates (primarily oxygen and water vapor), followed by creation of a damaged layer that exhibits graphitic bonding due to the high friction coefficient. When this surface is exposed to air, formation of the protective layer of adsorbates occurs and the friction coefficient is reduced once more. Removal of the damaged layer by wear and termination of the surface with oxygen may accompany the transition to low friction coefficient, just as occurred during the initial period of sliding in air. Although we have not performed surface analysis after the final period of sliding in laboratory air, the data obtained after sliding at relatively high water vapor pressure in test DD4 (Table VI) do indicate a return to more diamond bonding, in support of this hypothesis.

Gardos and Ravi *(11)* attributed an initial decrease in friction coefficient to the removal of adsorbed contaminants so that the friction behavior was representative of the CVD diamond surfaces. An opposite trend was observed in this investigation for identical specimens, but the vacuum conditions were significantly different. The data presented by Gardos and Ravi *(11)* were obtained in the relatively modest vacuum $(1.3 \times 10^{-3}$ Pa$)$ of a scanning electron microscope. Under the reciprocating contact conditions employed *(11)*, the monolayer adsorption time was much less than the time between passes of the slider. The initial decrease in friction when sliding was started under these conditions was therefore more likely attributable to a reduction in surface roughness than to desorption of surface contaminants. However, our friction data obtained in reduced pressures of oxygen and water vapor suggest that gas pressures greater than 10^{-3} Pa are required to produce the same low friction behavior as observed in air. Our vacuum test conditions $(< 6 \times 10^{-7}$ Pa$)$ result in a monolayer adsorption time on the order of 200 times greater than the time between passes of the slider (1.1 seconds). Our results are in agreement with other literature results for single crystal diamond rubbing against diamond in vacuum *(1-2, 7)*.

CVD diamond surfaces were found to contain up to several percent oxygen, even after sliding in ultrahigh vacuum (see Tables V and VI). This may be the result of adsorption from the vacuum system, since the electron beam parameters chosen to prevent damage to the diamond films resulted in relatively long analysis times. Alternatively, the oxygen may reside on non-contacting portions of the diamond asperities sampled simultaneously with the worn regions.

Implications for Chemical Analysis of Diamond Microcontacts. The principal difficulties associated with the chemical characterization of diamond tribological surfaces are due to the size of the contacts and the sensitivity of diamond to electron (and ion) beam induced degradation. Auger electron spectroscopy is capable of distinguishing diamond from other forms of carbon, and is probably the best surface science tool available for such characterization due to the amount of information obtained and the small size of the chemical probe. However, for microcrystalline films, even the smallest electron beam diameter may cause several different charging domains to be sampled simultaneously. Figure 2b indicates that with a 1.6 μm beam, we are sampling several asperities during analysis of the smooth diamond film, whereas with the larger asperity size on the rough diamond (Figure 3b) we are likely to probe individual asperities with the same electron beam. Statistically, the regions of actual contact will be smaller and farther apart than even the asperities shown in Figures 2 and 3. State-of-the-art Auger microprobes do not have sufficient spatial resolution to focus the probe on individual diamond contacts for most films, even in the absence of charging effects which will defocus the electron beam. A statistical approach is the obvious solution, but the time required to sample enough points on the surface to establish reasonable confidence in the data is prohibitively long, particularly when low beam current and pulse counting detection are required to prevent electron beam induced alteration of the diamond surface. More efficient instruments with multi-channel detection capabilities will alleviate these difficulties to some degree. It may also be possible to qualitatively assess the bonding character of worn diamond surfaces by scanning a large area of the wear track (relative to the asperity size), thereby sampling a large number of contact spots simultaneously.

Conclusions

It was found that the absence of oxygen and water vapor decreases the friction coefficient of 440C stainless steel in contact with CVD diamond. There are two possible sources of this reduction. In vacuum, the surface of the steel can not oxidize, so that diamond asperities plow through the metallic surface rather than penetrate an oxide which continually forms on the steel surface as sliding progresses in air. Secondly, direct contact of the diamond asperities by iron in vacuum may catalyse a transformation to graphite, in the same way that diamonds are polished on a cast iron wheel. This transformation may be suppressed by the formation of an oxide barrier at the surface of the steel in air.

The position of the C_{KVV} and plasmon peaks may not be useful when materials other than diamond slide against diamond, because the transfer and accumulation of foreign material on the diamond surface may lead to varying charging behavior and electron density independent of changes to the surface bonding state.

The friction of CVD diamond sliding on itself increases when oxygen and water vapor are removed from the environment. Removal of these surface films by abrasion and desorption allows carbon-carbon bonding across the interface, and perhaps the interaction of partially terminated surfaces. In either case, the adhesive interaction of the diamond surfaces results in an increase in the friction coefficient. Adsorption from the background environment of the vacuum system during long idle periods or controlled exposure of the surfaces to oxygen or water vapor at sufficiently high pressure can reduce the friction coefficient from the clean-surface value. Adsorption and desorption of gaseous species seem to have more effect on the friction that debris accumulation and removal.

For equivalent pressures, water vapor seems to be more effective than oxygen in reducing the friction coefficient of diamond sliding against diamond. This may be a result of the ability of the diamond surface to rapidly adsorb a complete layer of water vapor, even at low pressures, whereas the same surface may have a much lower sticking coefficient for molecular oxygen. Oxygen at much higher pressures may be capable of producing similar reduction in the friction coefficient.

The surface analysis data suggest that the diamond surface exhibits more sp^3 bonding when sliding against diamond in the presence of air than when sliding occurs in vacuum. This can be attributed to removal of adsorbed contamination from handling and any damage layers produced during the deposition process. The result is a relatively low friction coefficient of 0.1 or less. Likewise, sliding in vacuum environments causes a shift to more sp^2 bonding at the surface. This may be attributed to the creation of a damaged layer due to the high friction coefficient which occurs in vacuum and the resulting high surface shear stresses.

Due to the large variability in performance and chemistry observed on different areas of the diamond surface under identical test conditions, relative changes in friction coefficient and surface chemistry during sliding on a single wear track must be used to determine how reaction of the diamond, counterface material, and environment affect friction performance and surface chemistry.

Acknowledgments

The authors would like to express their appreciation to Elizabeth Sorroche for performing the friction measurements and surface analysis, and to Dave Tallant and Regina Simpson for performing the Raman spectroscopy. We would also like to thank Paul Hlava for useful discussions on collection techniques for natural diamonds. This work was partially supported by the Air Force Strategic Defense Initiative Organization under contract FY1457-91-N-5020.

Literature Cited

1. Bowden, F.P.; Young, J.E. *Proc. Roy. Soc.* **1951,** *A208,* pp 444-455.
2. Bowden, F.P.; Hanwell, A.E. *Proc. Roy. Soc.* **1966,** *A295,* pp 233-243.
3. Bowden, F.P.; Brooks, C.A. *Proc. Roy. Soc.* **1966,** *A295,* pp 244-258.
4. Miyoshi, K.; Buckley, D.H. *Appl. Surf. Sci.* **1980,** *6,* pp 161-172.
5. Tabor, D. In *The Properties of Diamond;* Field, J.E., Ed.; Academic Press: New York, NY, 1979; pp 325-350.
6. Seal, M. *Phil. Mag.* **1981,** *A43,* pp 587-594.
7. Hayward, I.P. *The Friction and Strength Properties of Diamond;* Ph.D. Thesis, University of Cambridge, 1987; pp 71-94.
8. Samuels, B.; Wilks, J. *J. Mat. Sci.* **1988,** *23,* pp 2846-2864.
9. Jahanmir, S.; Deckman, D.E.; Ives, L.K.; Feldman, A.; Farabaugh, E. *Wear* **1989,** *133,* pp 73-81.
10. Evans, T. In *The Properties of Diamond;* Field, J.E., Ed.; Academic Press: New York, NY, 1979; pp 403-424.
11. Gardos, M.N.; Ravi, K.V. "Tribological Behavior of CVD Diamond Films," Proc. 1st Int. Conf. on Diamond and Diamond-Like Films, Electrochemical Society, Los Angeles, CA, May 7-12, 1989, paper #115.
12. Blau, P.J.; Yust, C.S.; Heatherly, L.J.; Clausing, R.E. In *Mechanics of Coatings;* Dowson, D.; Taylor, C.M.; Godet, M. Eds.; Elsevier: New York, NY, 1990; pp 399-407.
13. Hayward, I.P.; Singer, I.L.; Seitzman, L.E. "Effect of Roughness and Surface Chemistry on Friction of CVD Diamond Films," presented at the 17th Int. Conf. on Metallurgical Coatings and 8th Int. Conf. on Thin Films, San Diego, CA, April 2-6, 1990.
14. Hayward, I.P. "Friction and Wear Properties of Diamonds and Diamond Coatings," presented at the 18th Int. Conf. on Metallurgical Coatings and 9th Int. Conf. on Thin Films, San Diego, CA, April 22-26, 1991.
15. Peebles, D.E.; Pope, L.E. In *Advances in Engineering Tribology;* Chung, Y.-W.; Cheng, H.S., Eds.; STLE Special Publication SP-31; The Society of Tribologists and Lubrication Engineers: Park Ridge, IL, 1990; pp 13-30.
16. Peebles, D.E.; Pope, L.E. *J. Mat. Res.* **1990,** *5,* pp 2589-2598.
17. McDonald, T.G.; Peebles, D.E.; Pope, L.E.; Peebles, H.C. *Rev. Sci. Instr.* **1987,** *58,* pp 593-597.
18. Li, F.; Lannin, J.S. *Phys. Rev. Lett.* **1990,** *65,* pp 1905-1908.
19. Wilks, J; Wilks, E.M. In *The Properties of Diamond;* Field, J.E., Ed.; Academic Press: New York, NY, 1979; pp 351-382.
20. Thomas, J.M. In *The Properties of Diamond;* Field, J.E., Ed.; Academic Press: New York, NY, 1979; pp 211-244.
21. Harrison, J.A.; Brenner, D.W.; White, C.T.; Colton, R.J. "Atomistic Mechanisms of Adhesion and Compression of Diamond Surfaces," presented at the 18th Int. Conf. on Metallurgical Coatings and 9th Int. Conf. on Thin Films, San Diego, CA, April 22-26, 1991.

RECEIVED November 26, 1991

Chapter 6

The Role of Sulfur in Modifying the Friction and Lubricity of Metal Surfaces

G. A. Somorjai and M. Salmeron

Department of Chemistry, University of California—Berkeley,
Berkeley, CA 94720
and
Materials Sciences Division, Lawrence Berkeley Laboratory,
Berkeley, CA 94720

Recent surface crystallography studies using scanning tunneling microscopy and low energy electron diffraction reveal the coverage dependent changes of the surface structure of sulfur on rhenium surfaces. There is evidence of metal reconstruction as the sulfur metal bonds are formed and there is evidence of clustering of sulfur atoms that form trimers, tetramers and hexamers at increased coverages. Contact experiments using the STM reveal that the elastic properties of the surface are maintained by the protecting S overlayer. The implications of the STM and LEED crystallography findings to tribology and the importance of sulfur monolayers in modifying the mechanical properties of surfaces are discussed.

In recent years the development of the scanning tunneling microscope and the atomic force microscope permits us to obtain real space images of surface atoms and adsorbed molecules and to apply forces in the range of 10^{-9}-10^{-4} Newtons on the atomic scale that are less or slightly more than required to break bonds between atoms. New non linear laser spectroscopy techniques are being developed (SHG and SFG) that are not only totally surface sensitive but provide access to the buried solid-solid or solid-liquid interfaces that are not accessible to electron or ion probes (1). Therefore, it is not surprising that the attention of surface scientists is now focusing on phenomena that occur at these solid-solid and solid-liquid interfaces that include mechanical properties such as adhesion, friction and lubrication. Our aim is to understand, on the molecular level, how bonding occurs and how the mechanical properties at these surfaces are determined by the surface structure and surface bonding.

To this end, we undertook studies of the passivating properties of sulfur monolayers on the Mo(001) and the Re(0001) crystal faces. Sulfur forms an ordered monolayer on these metal surfaces and exhibits surprising and diverse structural chemistry as a function of coverage. We studied these sulfur monolayer structures by both LEED and STM (2-6). The adsorbate induced restructuring of rhenium has been identified by LEED crystallography (7). Substrate mediated adsorbate-adsorbate (sulfur-sulfur) interactions gives rise to clustering in the otherwise already ordered sulfur layer to produce 3, 4 and 6 member sulfur clusters of unique geometry. Sulfur is an important component in many organic and inorganic lubricants. The overlayers that it forms have structural and chemical properties that illustrate many of the reasons for its unique lubricity and passivating characteristics against mechanical or chemical attack on surfaces.

Mechanical Force Needed to Break a Chemical Bond

Consider a surface atom bound to its neighbors by a chemical bond of binding energy of the order of 1eV. We can estimate the mechanical force, F, necessary to break this bond by computing the force needed to displace the atom by 1Å, the order of magnitude of bond lengths. Thus the force needed to stretch the bond is given by F = 1eV/1Å which yields 1.6×10^{-9} Newton. We need forces in the range of 10^{-9}-10^{-8}N per atom to break the bond which corresponds to 0.1-1μgm. The atomic force microscope can explore contact forces on the atomic scale. As the microscope tip approaches the surface at distances of 2Å or less, the forces (attractive then repulsive) exerted on the tip by the surface can be measured in the range as small as 10^{-9} Newton and up to 10^{-4} Newton. After applying a force of certain magnitude the area of tip contact is scanned for signs of permanent damage. Using a smooth Au surface (as shown in Figure 1), permanent damage could only be detected at about 10^{-5} Newtons with a tip radius of 10^3Å (8). This indicates that the metal surface responds to the approaching metal tip by elastically deforming in the range of up to 1.0 GPa pressure. This is the regime of elastic deformation. Only by applying a pressure greater than 1.0 GPa will plastic deformation commence on Au surfaces accompanied by irreversible displacement of Au atoms as shown in Figure 1. From measurements of this type one can determine the forces needed for elastic or plastic deformation on the atomic scale and to correlate the results with those obtained by macroscopic studies.

Sulfur Structures on Re(0001) Crystal Face.   LEED and STM Studies

The structures formed by sulphur on Re(0001) surfaces as a function of coverage θ, have been studied by LEED and AES (4). At saturation (θ=0.50) S imparts to the Re(0001) surface passivating properties. In particular it was shown that such a monolayer prevents oxidation of Re upon exposure to air. STM experiments revealed that the S-structure was stable and could be imaged in air (9). To study the structures formed at lower coverages, the experiments have to be performed in UHV. We have recently completed studies of the structures formed from around θ=0.25 up to θ=0.50 (7,8). The outcome of these studies constitutes a dramatic demonstration of how STM helps solve structures with large unit cells. The capability to study complicated structures is not only an incremental refinement of crystallographic techniques, but uncovers new phenomena, like new phases formed by adsorbate aggregation. We describe now briefly these findings. Up to a coverage of 0.25, S-S repulsive first neighbor pairwise interactions dominates the structure of the absorbed layers. The structure formed is (2x2) and is shown in Figure 2a . As the coverage increases, the next ordered structure of high symmetry that maintains the monomeric form of S is the ($\sqrt{3}$x$\sqrt{3}$)R30 when θ=0.33, as observed on many other single crystal surfaces (10,11). Instead, on Re(0001) a new phase forms in which S coalesces into trimers, first near domain boundaries and then at all regions. In these trimers, each sulphur atom sits on the same three-fold hollow site as in the low coverage monomer structures. The fact that dimers do not form except as occasional defects indicates that three-body forces are determinant at this stage. Three body forces were previously introduced as necessary corrections to the dominant pairwise interactions in Monte Carlo

Micro-Hardness of Gold
(2500Å gold on mica)

5mm=1000Å

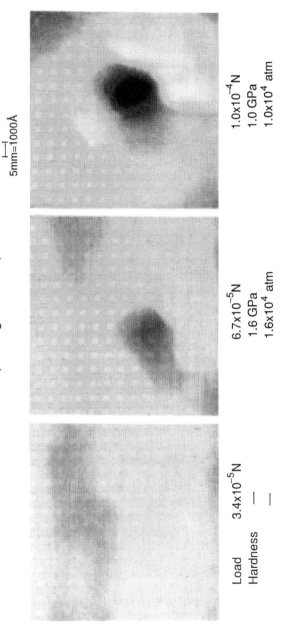

Load	3.4×10⁻⁵N	6.7×10⁻⁵N	1.0×10⁻⁴ N
Hardness	—	1.6 GPa	1.0 GPa
	—	1.6×10⁴ atm	1.0×10⁴ atm

Figure 1. Atomic Force Microscopy images of an oriented Au film on mica. Before each image the tungsten tip was pressed against the surface in the center of the image. Permanent indentations were only observed above 6.5×10^{-5} Newtons. The tip radius is estimated from the diameter of the indentation to be close to 1,000Å. After Blackman and Mate (8).

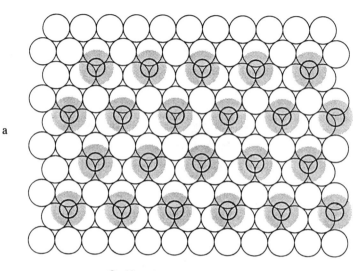

a

Sulfur 2 x 2 on Re(0001)

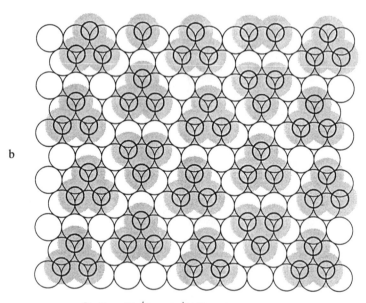

b

Sulfur (3√3 x 3√3)R30° on Re(0001)

Figure 2. Schematic diagram showing the S structures formed on Re(0001) as a function of coverage. The position of the S atoms (grey circles) was determined by LEED I-V analysis in the case of the (2x2) structure (a). (b), (c) and (d) show the various ordered S aggregate structures that were detected by STM. The adsorption site was determined by a combination of LEED and STM *(5,6,9)*.

c

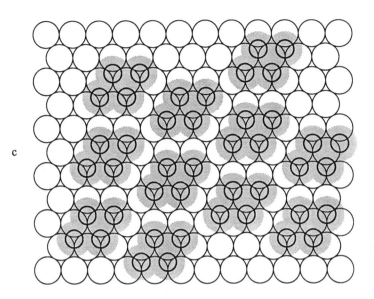

Sulfur ($\begin{smallmatrix}3&1\\1&3\end{smallmatrix}$) on Re(0001)

d

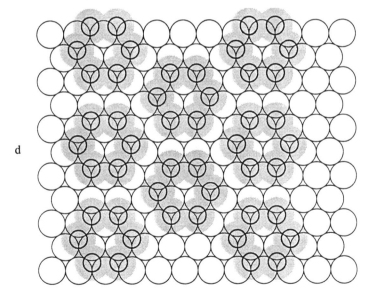

Sulfur (2√3 x 2√3)R30° on Re(0001)

Figure 2. Continued.

simulations of gas lattice dynamics in order to fit the experimental data (*12*). Here, these forces are dominant and cause the local coverage to change abruptly from 0.25 to near 0.45 (the value of θ for the ordered $(3\sqrt{3} \times 3\sqrt{3})R30°$ structure described below.) Initially, the trimers are identical and have their center in a three-fold hollow site. As the coverage continues to increase, trimers of a different structure form that are rotated 60° relative to the first ones and have their center on a top site. At maximum trimer density an ordered $(3\sqrt{3} \times 3\sqrt{3})R30°$ structure forms that is composed of trimers of the two types in a 3 to 1 ratio, as shown in Figure 2b. Atom counting indicates that the coverage is 0.45. By adding more sulphur to the surface, the trimers incorporate a fourth sulphur atom and rearrange to form tetramers. The schematic of Figure 2c shows the new structure formed at completion, a (3,1;1,3) in LEED matrix notation. The coverage, determined by counting atoms, is 0.5. This observation came as a surprise. The highest coverage structure (and saturation) was thought to be the one obtained by further sulfur dosing, with a $(2\sqrt{3} \times 2\sqrt{3})R30°$ structure, where S forms hexagonal units as shown in Figure 2d and in the image at the top of Figure 3. The coverage here is again 0.5. To form this last structure, a higher exposure to H_2S or S_2 is required. By heating this saturation structure to increasingly higher temperatures to desorb sulfur, all the lower coverage ones can be generated, including the tetrameres that coexist with the trimers and defects. The $(2\sqrt{3} \times 2\sqrt{3})R30°$, structure might be energetically more stable than the (3,1;1,3), but it requires a substantial rearrangement of sulphur atoms in order to break the trimer unit. Thus the tetrameres structure is kinetically stabilized against the more stable hexagonal ring structure.

The Role of Coadsorption and Strongly Adsorbed Monolayers During Tribological Change (Friction, Lubrication)

High coverages of chemisorbed molecules exhibit decreasing heats of adsorption indicating the weakening of the average adsorbate-substrate bond. This effect is due to the adsorbate-adsorbate interaction that often forces the molecule onto new adsorption sites. The coadsorption of two different molecules, one an acceptor the other a donor to the metal substrate, leads to ordering into a mixed layer due to attractive adsorbate-adsorbate interaction. One example of this is the surface structure that forms upon the coadsorption of carbon monoxide and ethylene (*13*). The coadsorption of two donors or two acceptors yields disordered monolayers as indicated by several studies (*14*). However, there are many examples indicating that one of the chemisorbed species during coadsorption can also restructure the substrate. Chemisorbed potassium restructures iron oxide (*15*), oxygen or alumina restructures iron (*16*) and sulfur or carbon restructures nickel and rhenium (*7*). Thus, the chemisorbed coadsorbed species exert their chemical influence not only by altering the adsorbate-adsorbate bond but also through restructuring of the substrate. This effect is particularly noticeable during catalytic reactions and could be very important during lubrication. A monolayer of strongly chemisorbed atoms, sulfur for example, can act as a lubricant markedly decreasing the friction coefficient of the surface. The good lubricating properties of S compounds derive in our opinion from the strong S-metal bonds formed at the surface. The strength of these bonds is such that S-covered surfaces are chemically passivated against chemical attack, even when the new bonds to be formed by the reactive species are stronger than those formed by S. The reason for this is kinetics, not thermodynamic equilibrium. At saturation S-coverages, oxidation by air exposure, for example, is inhibited by the lack of empty or exposed Re sites that are necessary to dissociate the attacking O_2 molecule.

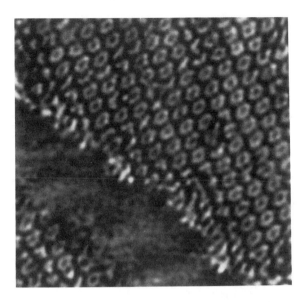

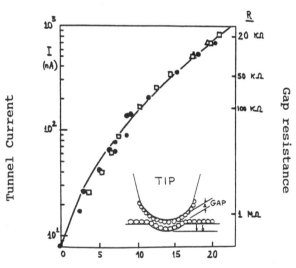

Figure 3. (Top) STM topographic image showing the $(2\sqrt{3} \times 2\sqrt{3})R30°$ sulfur structure. Hexagonal rings of sulfur constitute the unit cell and are clearly visible in the image, along with defects and a domain boundary running diagonally. The blurred region is a step, two atoms high, separating two terraces. The compression of the distance scale along the diagonal is the result of thermal drifts. (Bottom) Tunnel current and gap resistance as a function of tip displacement over a surface similar to that of the image shown on top. Tip displacement is measured relative to the initial tunneling conditions of 8 nA and -16mV sample bias. The symbols show different runs from the same sample and tip and the solid line is calculated by assuming the current increase is due solely to an increase of the area of contact for a conical tip.

During mechanical contact, S overlayers can thus prevent the formation of metal-metal bonds between the contacting bodies. The situation can be compared to that of layer compounds such as MoS_2. Cleavage along the Van der Waals layers, that are separated by 2 sulfur layers, is easy and does not disrupt bonds within the layers.

We have performed studies to illustrate these ideas by studying the elastic contact between a PtRh tip (probably covered with S) and a Re(0001) surface that is also saturated with S. We describe the results in the next section.

Elastic Properties of Sulfur Monolayers on Metals

In order to study the tip-surface contact, we measured the changes in tunnel current as a function of tip displacement over the sulfur covered Re(0001) surface (*17*). An ordered $(2\sqrt{3}x2\sqrt{3})R30°$ sulfur overlayer was first prepared in UHV and the sample was afterwards transferred through air to an STM operating in a 10^{-7} Torr vacuum.

In the experiments presented here, the $(2\sqrt{3}x2\sqrt{3})R30°$ sulfur structure was first imaged in the topographic mode at a gap resistance of $20M\Omega$ and a -16mV sample bias (Figure 3). The tip was then positioned in the middle of the image, the feedback loop was disabled, and the tip was advanced until a predetermined current I_{max} was obtained. Currents as large as 900 nA were obtained after z piezo displacements of 20Å, as shown in Figure 3. The gap resistance reached values as low as 20 kΩ, on the order of the Sharvin point contact resistance. However, no changes were seen in atomic resolution images of the sulfur overlayer recorded before and after tip approach, indicating that only elastic deformations of the tip or sample took place during the contact. Advances larger than 20Å and 900nA did however result in irreversible disruption of the surface.

Tip-surface forces could not be directly measured in these STM experiments. However, if we assume that most of the 20Å displacement is due to elastic deformations of tip or sample, we can get to an estimate. First, we assume a rigid gap and a deformable tip and/or surface with increasing area of contact as the tip advances. The increase in current would then be proportional to the increase in contact area and we can treat the measured gap resistance as the Sharvin resistance $R = 4\rho l/3\pi a^2$. Here ρ is the specific resistivity of the material, l is the electron mean free path, and $a = a(z)$ is the radius of the circular contact area as a function of tip displacement z. It depends on the tip shape as the only parameter. Indeed, we obtain a good fit to the data of Figure 3 by implying a conical tip (solid line). Using the elastic moduli of Re and PtRh and a radius of 100Å for the tip (estimated from the observed step widths in the STM image), the initial compressive force is estimated to be of the order of 10^{-6} N and the maximum gap pressure $\approx$10GPa.

Summary

The combination of surface structural studies on well characterized surfaces using LEED and STM and measurements of the mechanical properties by AFM appear promising for studies of tribological properties (friction, lubrication) on the molecular scale. There are many surprises along the way as demonstrated by studies of the S/Re(0001) system. There is ordering in the chemisorbed S layer, chemisorption induced restructuring and complex adsorbate-adsorbate interaction among the sulfur atoms leading to clustering (S_3, S_4 and S_6). Many of the properties of the S/Re(0001) interface system identified in these investigations and in particular the very strong S-Re bonds that allow effective blocking of Re sites to further chemical attack, contribute to make sulfur a premier ingredient in lubricants and lubricating systems.

Acknowledgment

This work was supported by the Director, Office of Energy Research, Office of Basic Energy Sciences, Material Sciences Division, U.S. Department of Energy under contract No. DE-AC03--76SF00098.

Literature Cited

1. Shen, Y.R. In *Chemistry and Structure at Interfaces*; Editor, R.B. Hall and A.B. Ellis; VCH, Florida, 1986; pp 151-196.
2. Salmeron, M and Somorjai, G.A. *Surf. Sci.* 1983, *127*, 526.
3. Marchon, B.; Bernhardt, P; Bussell, M.E.; Somorjai, G.A.; Salmeron, M. and Siekhaus, W. *Phys. Rev. Letters* 1988, *60*, 1166.
4. Kelly, D.G.; Gellman, A.J.; Salmeron, M.; Somorjai, G.A.; Maurice, V. and Oudar J. *Surf. Sci.* 1988, *204*. 1.
5. Hwang, R.W.; Zeglinski, D.M.; López Vázquez-de-Parga, A.; Ocal, C.; Ogletree, D.F.; Somorjai, G.A.; Salmeron, M. and Denley, D.R. *Phys. Rev. B.* 1991, *44*.
6. Ogletree, D.F.; Hwang, R.Q.; Zeglinski, D.M.; López Vázquez-de-Parga, A.; Somorjai, G.A. and Salmeron, M. *J. Vac. Sci. Technol. B* 1991, *9*, 886.
7. Jenz, D.;Held, G.; Wander, A.;Van Hove, M.A. and Somorjai, G.A. (to be published).
8. Blackman, G.; Mate, M.; Somorjai, G.A. unpublished results.
9. Ogletree, D.F.; Ocal, C.; Marchon, B.; Salmeron, M.; Somorjai, G.A.; Beebe, T.P. and Siekhaus, W. *J. Vac. Sci. Technol. A* 1990, *8*, 297.
10. Ohtani, H.; Kao, C.-T.; Van Hove, M.A. and Somorjai, G.A. *Progress in Surface Science* 1986, *23*, 155.
11. Van Hove, M.A.; Wang, S.W.; Ogletree, D.F. and Somorjai, G.A. *Advances in Quantum Chemistry* 1989, *20*, 1.
12. Einstein, T.L. *Surf. Sci.* 1979, *L497*.
13. Blackman, G.S.; Kao, C.-T.; Bent, B.E.; Mate, C.M.; Van Hove, M.A. and Somorjai, G.A. *Surf. Sci.* 1988 *207*, 66.
14. Mate, C.M.; Kao, C.-T. and Somorjai, G.A. *Surf. Sci.* 1988 *206*, 145.
15. Vurens, G.H.; Strongin, D.R.; Salmeron, M. and Somorjai, G.A. *Surf. Sci. Letters* 1988 *199*. L387.
16. Strongin, D.R.; Bare, S.F. and Somorjai, G.A. *J. Catal.* 1987 *103*, 289.
17. Salmeron, M.; Ogletree, D.F.; Ocal, C.; Wang, H.-C.; Neubauer, G. and Kolbe, W. *J. Vac. Sci. Technol. A*, (in press)

RECEIVED October 21, 1991

Chapter 7

Tribology of Ceramic and Metallic Surfaces in Environments of Carbonaceous Gases

Application of Raman and Infrared Spectroscopies

James L. Lauer and Brian L. Vlcek

Department of Mechanical Engineering, Aeronautical Engineering and Mechanics, Rensselaer Polytechnic Institute, Troy, NY 12180–3590

Friction and wear properties of carbons and cokes overcoating ceramic and metallic wear surfaces were determined over a range of temperatures up to 700°C and pressures up to 7 GPa. In order to correlate the mechanical properties of the carbons with chemical properties, such as bonding, and with crystal habits, pertinent literature on Raman and infrared spectroscopies of both carbons and ceramic substrates used as tribosurfaces has been reviewed. All the tribological data were measured with a pin-on-disc tribometer on carbon generated by the dehydrogenation and pyrolysis of ethylene on steel or ceramic substrates, such as Si_3N_4, SiC, $A\ell_2O_3$, or sialon, as well as on the bare substrates themselves. Reference is also made to pertinent parts of the now already voluminous literature on the vibrational spectroscopies and tribologies of the head/magnetic media interface in computer applications.

For a number of years our laboratory has been engaged in investigations on the use of carbonaceous gases as feeds to supply lubricants to bearing and other moving surfaces operating at temperatures above 350°C. As internal combustion engines are becoming smaller and lighter to reduce fuel consumption, their operating temperatures increase to the point where water-cooling is no longer practical. Doing away with the cooling system further reduces engine weight and increases temperatures above the tolerance of liquid lubricants. Solid lubricants such as graphites and other forms of lubricating carbon have been shown to reduce friction and wear at high temperatures but are rapidly consumed and require replenishment. If supplied as powders their effectiveness is much less than when they are formed by surface reactions with gaseous feeds because of poor adhesion. Experience in petroleum cracking and refining has shown that carbon - or coke-forming reactions [carbon and coke are used loosely to mean materials primarily consisting of elemental carbon but also containing small quantities of other elements, mainly hydrogen] will occur on transition-metal surfaces, e.g. on nickel or palladium-containing alloys even at temperatures below 150° and on some silica-alumina clays at around 300°C or somewhat less but it remained for us to demonstrate that some of these carbons can be very effective

0097–6156/92/0485–0112$08.75/0

lubricants [*1*]. Klaus and coworkers have been using phosphoric acid esters to form lubricating reaction products at wear surfaces [*2*]. Though less reactive and therefore slower in forming, the carbon feeds consume less substrate material and have the potential of eliminating the need for a separate lubricant supply since part of the fuel itself or exhaust gases can be the reagents.

The type of lubrication suggested for these applications is a boundary lubrication where the lubricant can no longer be supplied as an additive in a bulk liquid but is surface-generated by reaction with a continuously supplied gas. Preferably this gas should be thermally stable enough not to react prior to its contact with the moving surfaces and not require heated ducts to prevent condensation. Carbonaceous gases can be very stable thermally to meet this requirement while phosphoric acid esters can form highly corrosive phosphoric acids by condensation. On the other hand, most carbonaceous gases are flammable in certain proportions with oxygen while phosphoric acid esters are not. Oxygen was excluded in the experiments described in this paper. Boundary lubrication by any surface-interactive feed will naturally depend on the properties of the adsorbing substrate which is preferably a ceramic at temperatures much above 300°C. Dehydrogenation to a coke proceeds non-catalytically on most ceramic surfaces and therefore rather slowly. Some ceramic surfaces are exceptional and do function already at 300°C or even below to deposit carbon, but most ceramic surfaces must be at higher temperatures for pyrolytic carbon deposition.

Graphitic carbon formed on metal surfaces, carbons or cokes on activated silica-alumina, and pyrolytic carbons (like cokes also mostly hydrocarbons of low carbon content) on other ceramics differ in their tribological properties as well as in their rates of formation. The different tribological properties of the carbons or cokes are expected to be related to their chemical nature, their physical nature, rates of formation and their substrates, which sometimes are chemically and physically changed by carbon deposition and wear. The correlation of such data has been the subject of our ongoing research [*3*]. Raman and infrared spectroscopies play an outstanding role in the data analysis. Superior lubrication procedures are expected to result from this effort.

Many types of carbon have been known for a long time, graphite and diamond being the extremes of a long series that includes glassy carbon, several types of pyrolytic carbons [*4*], graphitizable and non-graphitizable carbons, carbonados, coals and cokes, diamond-like amorphous carbons and many others. Graphite rods are being used as nuclear moderators; electrode carbons and carbon resistors have been used in electrical applications. Carbon fibers are used to reinforce plastics and even metals such as aluminum. However, Raman and infrared spectroscopies are only now becoming common tools for carbon characterization. Carbon coatings for tribological applications are a recent development [*5*]. Diamond films grown at low temperatures and pressures have become common recently, although patents disclosing such processes were published in the 60's by Eversole [*6*] and by Lauer [*7*].

The subsequent sections will show the applicability of Raman and infrared spectroscopies to tribological surfaces in general but concentrate on surfaces overlaid by carbon. Since hard disc magnetic storage media have been an important tribological application for carbon coatings, some reference will be made to them. Then our work on metallic and ceramic tribosurfaces lubricated by surface-generated carbon will be described and characteristic Raman spectra shown.

REVIEW OF INFRARED AND RAMAN SPECTRA OF WEAR SURFACES

Wear surfaces of metal or metal oxides do not lend themselves to analysis by infrared or Raman spectroscopies. However, whenever the metal or metal oxide interacts with a liquid or gaseous lubricant or a polymeric countersurface, these spectroscopies can be applied to the surface deposits formed. In general, these deposits are not good lubricants and are not intentionally formed, our own work described in this article being an exception. Non-metals are, however, readily analyzed by these spectroscopies. In some applications, non-metals are applied to metallic magnetic storage surfaces as coatings, e.g. carbon may be sputtered on aluminum-based hard computer discs to increase their wear resistance [5]. The spectrum of these carbon surfaces are analogous to those of the deposit carbons except for their general lack of bands associated with bonding to their substrates.

The following brief literature survey is divided into sections of not carbon-coated and carbon-coated wear surfaces. It is necessary to include both types of surfaces to provide background for the description of our own work which involves chemical interactions of non-lubricating gas with a surface to form a carbonaceous coating.

(a) Not Carbon-Coated Surfaces

Considerable literature exists on the thin and usually brownish deposits which are formed within or very close to the wear track on metals and sometimes also on non-metals whenever organic materials are present and which have become known as "Friction Polymers." One of us [8] wrote a review of them some years ago. Friction polymers have been credited with promotion of good friction and wear characteristics at low and medium temperatures and authors, especially Furey [9], have suggested the inclusion of polymer-forming monomers in the liquid lubricant to enhance friction polymer formation. Figure 1 shows a recently published micro-FTIR spectrum of brown film deposits on a metal surface within and without the wear track [10]. The most pronounced difference between the spectra is the broad band near 1070 cm^{-1} in the wear track spectrum, which has been attributed to the presence of a C-O-C linkage, probably ether. The absence of a carbonyl band (C=O), usually near 1730 cm^{-1} or so, is noteworthy, because it points to formation of the film by a pyrolysis mechanism similar to that of coal formation. Small concentrations of oxygen must be present for this deposit to be formed.

At temperatures above 350°C ceramic wear surfaces are more appropriate than those of metals. Because of its hardness silicon carbide is often used. There exists generally a layer of silicon oxides and hydroxides in wear tracks formed in the atmosphere. Micro-FTIR is an appropriate technique for surface analysis because the wear tracks are so narrow. Figure 2 is such a spectrum from an unlubricated wear test [10]. The broad band near 3500 cm^{-1} is characteristic of hydrogen-bonded OH while the narrow band near 700 cm^{-1} is characteristic of Si-OH. The doublet near 1200 cm^{-1} is attributable to Si-O and the weak bands next to it at higher frequencies are attributable to amorphous carbon. It is likely that hydrated silicates were formed. These silicates can be a lubricating form of "water glass," which is a viscous liquid. These bands are found only in the wear track. The infrared spectra show evidence of carbon but only a weak one. Raman spectra are much superior for this purpose.

Raman spectra of SiC wear tracks also show bands not present outside the wear track, in particular, at 520, 1000, 1350 and 1600 cm^{-1} (Figure 3) [11]. The 520 cm^{-1} band is very sharp and characteristic of elemental Si, while the 1350 and 1600 cm^{-1} bands are quite broad and resemble those of the carbon overlays on computer discs (Figure 20). The band at 1000 cm^{-1} is of intermediate width and is very likely due to C-O-C linkages. This band is analogous to the 1070 cm^{-1} infrared band of Figure 1. Therefore SiC is partially decomposed during friction under high loads. The "tribochemically" formed carbon could be the nucleus for further carbon deposition (*vide infra*).

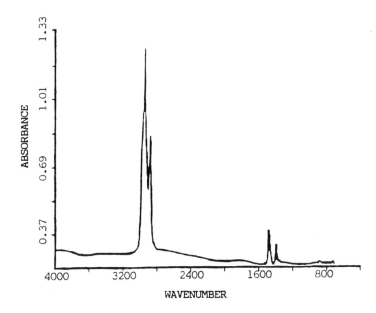

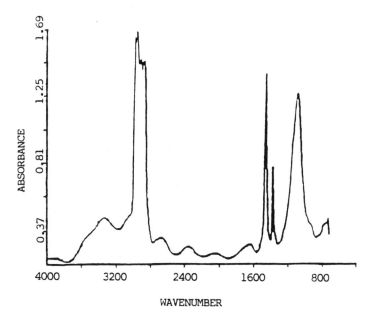

Figure 1. FTIR spectrum of an oil-lubricated wear surface outside (upper trace) and within the contact area (lower trace). [Reproduced with permission from reference [10], Hegemann, B.E., Jahanmir, S., and Hsu, S.M., "Microspectroscopy Applications in Tribology," *Microbeam Analysis* ,23, 193-195 (1988) (D.F. Newbury, ed).]

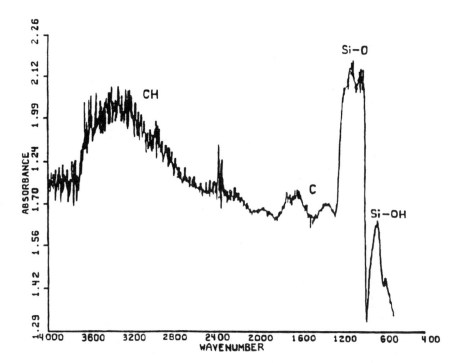

Figure 2. FTIR spectrum of the wear track of an unlubricated SiC-on-SiC friction couple. [Reproduced with permission from reference [10], Hegemann, B.E., Jahanmir, S., and Hsu, S.M., "Microspectroscopy Applications in Tribology," *Microbeam Analysis*, 23, 193-195 (1988) (D.F. Newbury, ed).]

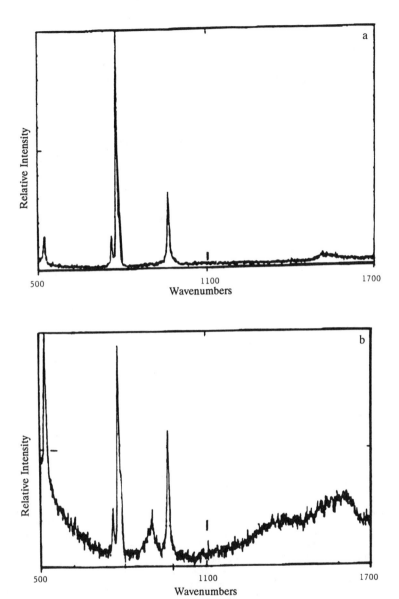

Figure 3.   Raman spectra of a SiC disc run against a SiC pin under high load (a) outside the wear track, (b) inside the wear track. [Reproduced with permission from reference [*11*], Exarhos, G. and Donley, M.S., "Real-Time Raman Detection of Molecular Changes in Ceramics," *Microbeam Analysis 22,* 125-127 (1987) (R.H. Geiss, ed).]

(b) **Carbon-Coated and Bulk Carbon Surfaces**

Carbon-coated or carbon wear surfaces have been used for a long time, e.g.in brushes of electric motors, in lead pencils whose "lead" is basically graphite, and in carbon paper and typewriter ribbons. However only recently have wear surfaces been studied by infrared and Raman spectroscopies because carbon surfaces have become important as overcoats on hard thin film magnetic storage discs and because of our own investigations of surface-generated lubricating carbon. Carbon, both elemental and in compounds with minor impurity elements, such as H, N, or O, exists in a very large number of forms. These forms depend on the method of formation. According to Kinoshita [12] elemental carbon itself occurs in three common forms--diamond, graphite, and amorphous carbon. Amorphous carbon has become of much interest lately because of its high-tech tribological applications. It may be considered as sections of the familiar hexagonal layers of graphite, in varying sizes and with little order parallel to the layers. When amorphous carbon is generated in structures of entangled fibers with significant hollow space between them, the resulting material can exhibit isotropic bulk properties and be almost as hard as diamond. ("diamond-like carbon").

Knight and White [13] recently compiled Raman spectra of graphitic and amorphous carbons and of different diamonds. Raman scattering comes from a skin depth of only 7 nm and yet produces good spectra, which is contrary to classical expectations. The reason is the absorption of the visible or near infrared excitation radiation, which produces a strong resonance effect enhancing the usually weak Raman scattered radiation by up to six orders of magnitude. Therefore the Raman cross-section of (black) graphite is much greater than that of (colorless) diamond. Figures 4, 5 and 6 show characteristic spectra obtained by these authors [13]. The spectra of the diamond films are particularly interesting because they have been used to show that diamond can be deposited in thin layers on surfaces as a result of gas phase pyrolysis of organic compounds. This low-pressure diamond synthesis is now routinely used. Thicker deposits are graphitic or amorphous [14].

There is extensive literature on Raman spectra of graphitized carbons, sputtered and annealed diamond-like carbon (DLC) and sputtered amorphous carbon. Since the purpose of this section is to serve as an introduction to the analyses used for tribosurfaces and not as an extensive review of carbons, only a few pertinent spectra are shown. Thus Figure 7a associates the 1360 cm-1 Raman band with graphite crystal size and the 1580 cm-1 band with an in-plane vibration of the graphite crystal [15]. These two bands are the so-called D ad G bands, originally meaning diamond and graphite bands. Because of their association with Pauling's sp3 and sp2 orbitals, they are often designated in that way. Other interpretations exist, e.g. the association of the 1360 cm-1 band with graphite edge planes [16]. Grinding graphitic crystals to smaller size enhances the intensity of the D band. In amorphous carbons annealing to higher temperatures enhances the D band relative to the G band (Figure 7b) and there are simultaneous frequency shifts. The reasons for these changes are still under active investigation [17]. A study of the relation of the D and G band widths and peak separations of amorphous carbon overcoats to wear of hard discs was published recently [18]. It will be described in some length in the Discussion.

Infrared spectra have been used less commonly than Raman spectra to identify carbon structures. Their application has been mainly to characterize C-H functional groups in diamond-like carbon films. For this purpose the carbon overcoats were removed from their substrates and transmission spectra obtained. Reflection spectra have also been obtained by the author and others. The selection rules for DLC amorphous carbons and diamonds films would exclude the D and G bands from the infrared spectrum, yet they have been found in DLC films after annealing to different temperatures (Figure 8) [19]. Their relative intensities move in opposite directions

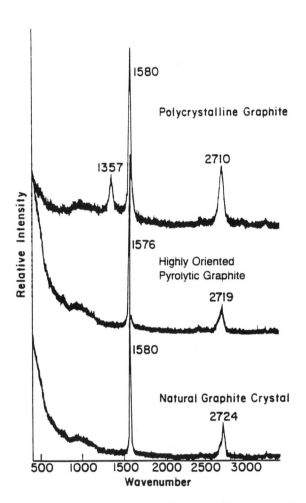

Figure 4.   Raman spectra of various crystalline graphitic carbons. [Reproduced with permission from reference[*13*], Knight, D.S., and White, W.B., "Characterization of Diamond Films by Raman Spectroscopy," *J. Mater. Res.* *4*(2), 385-393 (1989).]

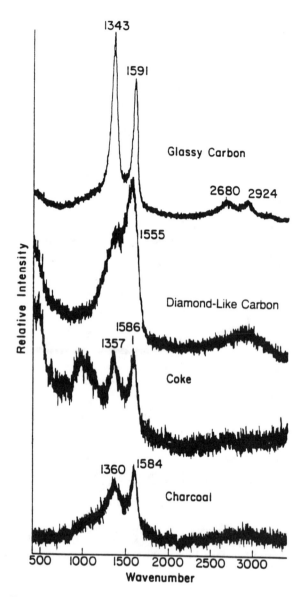

Figure 5. Raman spectra of various amorphous carbons. [Reproduced with permission from reference[13], Knight, D.S., and White, W.B., "Characterization of Diamond Films by Raman Spectroscopy," *J. Mater. Res.* 4(2), 385-393 (1989).]

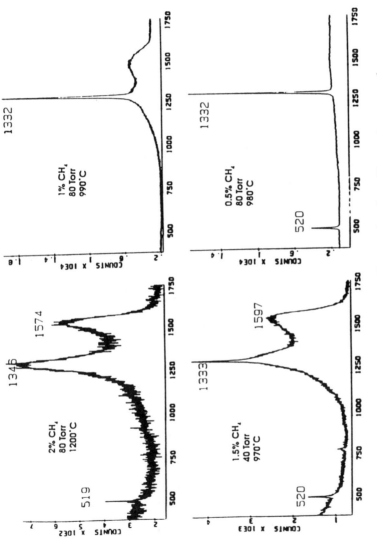

Figure 6.   Raman spectra of diamond films produced on silicon substrates under various conditions of deposition. [Reproduced with permission from reference[13], Knight, D.S., and White, W.B., "Characterization of Diamond Films by Raman Spectroscopy," *J. Mater. Res.* 4(2), 385-393 (1989).]

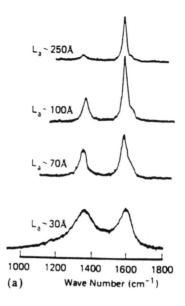

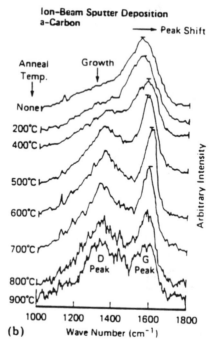

Figure 7.   Raman spectra of (a) graphitized glassy carbons of different crystal size and (b) amorphous carbon annealed at different temperatures. [Reproduced with permission from reference [17], Seki, H., "Raman Spectroscopy of Carbon Overcoats for Magnatic Disks," *Surface and Coatings Technology, 37*, 161-178 (1989).]

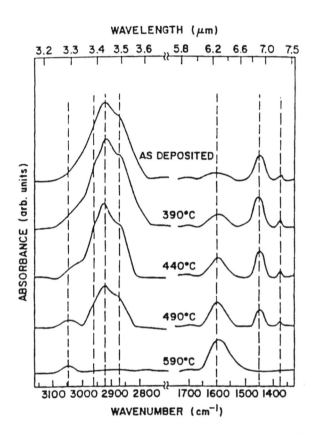

Figure 8.   IR spectra of DLC annealed at different temperatures. (The IR bands are listed and assigned in Table I). [Reproduced with permission from reference [*19*], Grill, A., Patel V., and Meyerson, B.S., "Optical and Tribological Properties of Heat-Treated Diamond-like Carbon," *J. Mater. Res.,* 5(11), 2531-2537 (1990).]

with temperature (The 1370 cm^{-1} band was assigned to C-CH$_3$ deformation and the 1600 cm^{-1} band to conjugated C=C). Infrared spectra published for coal particles can also show these bands in some instances (Figure 9), presumably depending on the conditions of coal formation [20]. Table I, taken from Reference 19 lists infrared bands of diamond-like carbon and their assignments

TABLE I
IR ABSORPTION FREQUENCIES OF DIAMOND-LIKE CARBON

WAVENUMBER (cm^{-1})	ASSIGNMENT	TYPE
3045	sp^2 CH	Stretching
2960	sp^3 CH$_2$ asymm.	Stretching
2920	sp^3 CH$_2$	Stretching
2875	sp3 CH3 symm.	Stretching
1600	sp2 C=C conjugated	Stretching
1450	C-CH3 asymm.	Deformation
1370	C-CH3 symm.	Deformation

[Reproduced with permission from reference [19], Grill, A., Patel V., and Meyerson, B.S., "Optical and Tribological Properties of Heat-Treated Diamond-like Carbon," J. Mater. Res., 5(11), 2531-2537 (1990).]

LUBRICATING CARBON FORMATION FROM CARBONACEOUS GASES AT METALLIC AND CERAMIC SURFACES

The work described below was carried out in our laboratory and some of it is published here for the first time. The original motivation was to use the coke generated as byproduct in catalytic petroleum refining as a solid lubricant directly on its support. Transition metal surfaces, such as nickel-containing alloys appeared to be most promising since they have been known to be active even at barely elevated temperatures in the formation of carbon [21]. Surprisingly, the reaction on nickel requires the presence of adsorbed hydrogen. According to Ref. (22), the mechanism is as follows:

$$C_2H_4 \Leftrightarrow CH(adsorbed) + H(adsorbed) \Leftrightarrow C(dissolved) \Leftrightarrow C(adsorbed)$$

$$\text{or} \qquad\qquad \Leftrightarrow Ni_3C(solid) \Leftrightarrow C'(adsorbed)$$

where C and C' are different forms of carbon. According to Lacava and coworkers [23], hydrogenated intermediates can be formed, specifically from benzene, but very likely also from olefins, and these intermediates in turn decompose to form carbon atmos which dissolve and diffuse through the metal to form carbon deposits or filaments. Our initial experimental work was carried out with a pin-on-disc tribometer at contact pressures of around 200 MPa and mostly with a sapphire pin on a nickel-coated sapphire disc rather than on pure metal [24]. This procedure allowed us to vary the coating while maintaining constant rigidity, as well as to compare the surface reaction rates with those of pure nickel films. The tribometer experiments are described in detail in the same article. For purposes of continuity it should suffice to state that a loaded stationary pin was in contact with a rotating disc while the contact region was heated by radiation from a projection lamp focused into it. A sliding thermocouple near the contact region served both as a temperature indicator and a temperature control sensor. Ethylene or other gases and vapors could be injected directly into the contact region from the open end of a stainless steel capillary tube connected to the gas supply. The surrounding atmosphere could be air or argon, care being taken in the former case to avoid flammable gas compositions near the ethylene

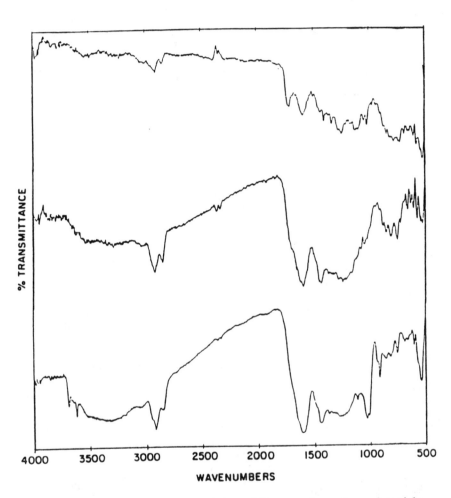

Figure 9.   IR transmission spectra of three different regions of a coal particle. [Reproduced with permission from reference [20], Krishnan, K., and Hill, S.L., "FTIR Microsampling Techniques."   Chapter 3 in *Practical Fourier Transform Spectroscopy*.   Academic Press, New York, NY 1990, pp. 103-165 (See in particular p. 121, Fig. 11).]

nozzle. Ethylene and the other gases used were applied at essentially ambient temperature to the contact region maintained at various known elevated temperatures; the actual contact temperatures were higher because of the frictional heat generated and could be calculated from theory, but were not measured. The friction coefficients were recorded throughout the experiment, i.e. at the controlled temperatures and in the presence or absence of gas injection; however, the analyses of the surface deposits were performed on the pins and discs (within and without the wear tracks) afterwards; i.e. at ambient temperatures and pressures. The very rapid cooling of the contact region insured irreversibility of the chemical changes that occurred during friction and wear at the higher temperatures. Figure 10 shows the changes of friction coefficient with time from the injection of ethylene into the contact region. As can be seen, the drop of friction coefficient became steeper as the temperature was raised until it became erratic above 650°C. Below 400°C a friction polymer might have been responsible for the decrease of friction, but above 400°C graphite formation will lowered the friction coefficient and dissolution of carbon in the nickel substrate might account for the erratic behavior about 650°C.

Auger electron spectroscopy of nickel films exposed to ethylene gas correlated very nicely with these results; steady carbon coverage developed over the same active temperature range (Figure 11) [25]. It is important to note from both Figures 10 and 11 that low friction and carbon coverage can be obtained already at 300°C.

In the next-higher temperature range a sapphire (alumina) pin was used on a Sialon (silicon nitride/aluminum nitride) disc in the presence of injected ethylene (Sialon always has a silica overcoat). The process of friction reduction worked in stages; once friction reduction was obtained, the next ethylene injection would produce a much faster friction drop (Figure 12) [24]. Benzene vapor could be substituted for ethylene with analogous results. Once a fast drop of the friction coefficient to a minimum value was reached, the experiment could be terminated and its resumption at a later time would bring the same drop of friction coefficient. This behavior provided a cogent argument for a mechanism of nucleation during "break-in." Furthermore a band at ~800 cm^{-1} was found in the Raman spectrum of a wear track of the Sialon disc after a tribometer run at 500°C with ethylene injection. This band is characteristic of SiC. The standard preparation of SiC (carborundum) is by reduction of SiO_2 with coal [26]. SiO_2 is always found on the surface of Si_3N_4 or Sialon, because of oxidation in air. It was already mentioned in the literature section above that SiC is partially decomposed into the elements under friction and wear even without the injection of ethylene or other carbonaceous gas into the conjunction region. However, the sharp friction reduction occurred only in the presence of ethylene. Hence it would seem reasonable to assume that carbon formed in the wear track by SiC dissociation might have provided the nuclei for further carbon formation--and thus lubrication--by carbon formed from the carbonaceous gas. However, this point is still under active investigation. Clearly, an intermediate is likely to be formed, which provides the nuclei for lubricating carbon deposition.

Auger and Raman spectroscopies have been the principal analytical tools to prove the presence of carbon in the wear tracks. Figure 13 shows significant differences in the Raman spectra for carbon produced on a metal (Ni_3Al) and on a ceramic (Sialon) [24]. Both carbon bands are sharper for the deposit formed on the metal than on the ceramic and the G peak is displaced toward lower wavenumber. The intensity ratios of D and G peaks are also different. These differences would support the presence of a graphitic deposit on the metal and an amorphous one on the ceramic. The formation of graphite deposits by catalytic dehydrogenation of hydrocarbons has been reported in the literature [27].

In our most recent work contact pressures up to 7 GPa were applied at bulk temperatures around 500°C. For this purpose the pin-on-disc apparatus was modified to use 0.8 mm diameter Si_3N_4 balls as pins to be run on Si_3N_4 plates. At linear

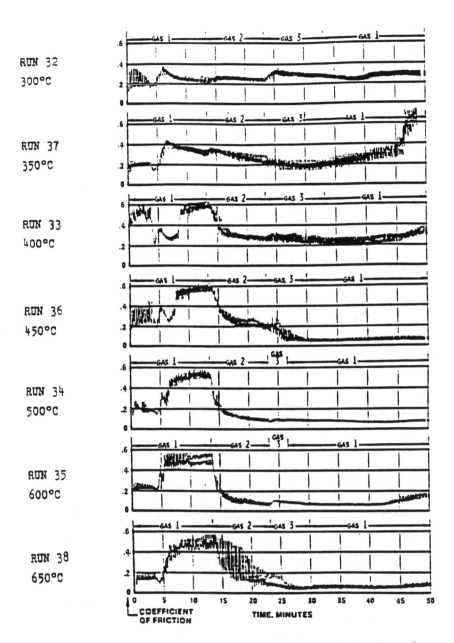

Figure 10.   Changes of friction coefficient with temperature in the pin-on-disc tribometer. (Nickel-coated Sialon disc, sapphire pin. Gas 1 = He+Ar, Gas 2 = C_2H_4 + Ar, Gas 3 = C_2H_4 + H_2+Ar). [Reproduced with permission from reference [1], Lauer, J.L., and Bunting, B.G., "High Temperature Solid Lubrication by Catalytically Generated Carbon," *Tribology Transactions 31*, 338-349 (1988).]

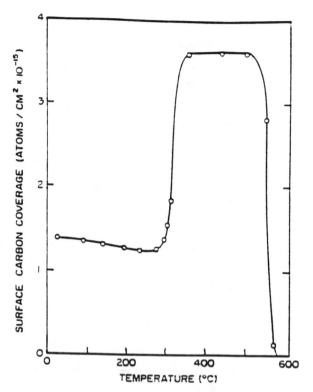

Figure 11.   Steady state surface carbon coverage, as determined by AES, at various substrate temperatures [25].

- TRIBOCHEMICAL WEAR-IN

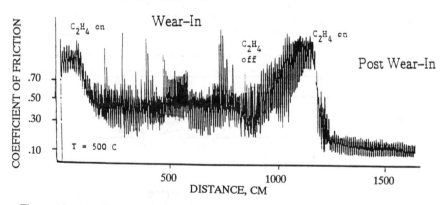

Figure 12.   Coefficient of friction vs. sliding distance for sapphire on bare Sialon. The ethylene reaction produces a much lower coefficient of friction after the wear-in period. [Reproduced with permission from reference [24], Lauer, J.L. and Dwyer, S.R., "Continuous High Temperature Lubrication of Ceramics by Carbon Generated Catalytically from Hydrocarbon Gases," *Tribology Transactions* 33(4), 529-534 (1990).]

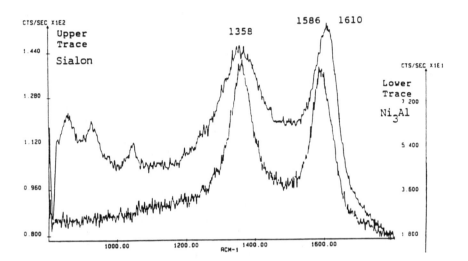

Figure 13.   Comparison of Raman spectra of surface carbon produced by ethylene on a metal (Ni3Aℓ) and a ceramic (Sialon).   [Reproduced with permission from reference [24], Lauer, J.L. and Dwyer, S.R., "Continuous High Temperature Lubrication of Ceramics by Carbon Generated Catalytically from Hydrocarbon Gases," *Tribology Transactions 33*(4), 529-534 (1990).]

speeds of around 10 m/s the flash temperatures were estimated to be in the 700-1000°C range. At these temperatures pyrolytic carbons should be readily deposited. Kinney described them in detail many years ago [4]. He collected a number of distinguishable types of carbon deposits by pyrolyzing different organic vapors in a quartz tube and labelled them A,B,C,D, and E. He also distinguished between organic vapors that produced graphitizable and non-graphitizable carbonaceous deposits.

Our recent work with pins and plates at high temperatures in the presence of ethylene produced carbonaceous deposits similar in appearance to those of Kinney and Eversole [6] Their friction data and analyses by Raman spectroscopy showed some interesting correlations. Figure 14 is a summary of the Raman spectra of the different carbons found in the wear track of a Si_3N_4/Sialon pin/disc experiment as well as on the pin and outside the wear track. The spectrum of the carbon on the pin shows a stronger D peak than G peak, which has been found to be generally true. The bottom spectrum of carbon just outside the wear track is characteristic of carbon black. The carbon could have originated within the wear track but been pushed aside by the pin, for that was the only carbon outside the wear track.

Figure 15 compares Raman spectra of carbon generated on silicon nitride and silicon carbide by a steel pin under identical conditions. The spectra are significantly different. All the Raman spectra obtained from the pins of the different friction couples shown in Figure 16 have higher D peaks than G peaks. Since the pins are continuously in the friction contacts while the plates are under stress only part of the time, the increased intensity of the D peaks could be the results of grinding down the crystals to small sizes as described by Lespade [15]. However, Lespade's work was with graphite while most of the carbons generated for Figure 16 were amorphous pyrolytic carbons. Perhaps the increased strength of the D band is also an indication of increased graphitization.

The wear track carbon spectra generated by a steel pin on silicon nitride (Figure 17) are all similar except for the carbon spectrum of the pin which shows the stronger D peak in agreement with our previous observations. Yet visually some of the carbons were very different. The narrow G band has been found to be characteristic of the carbon spectrum whenever a steel pin was present or when a sapphire pin was sliding on Sialon (Figure 18). The nature of the carbons produced under conditions of pyrolysis is not yet understood. This subject is now under active investigation by the authors and by other researchers. However, the Raman spectra are characteristic of operating conditions and have been found useful for identification.

RECENT FRICTION AND WEAR MEASUREMENTS ON CERAMICS

Figure 19 compares pin-on-disc friction changes with time under identical conditions for silicon nitride against silicon nitride without and with ethylene flow. Without ethylene flow the friction coefficient remains essentially constant while it first increases, then stays at a plateau and then decreases when ethylene is supplied. When the flow of ethylene is stopped at a time of low friction but the temperature is maintained and the flow in then resumed after some time, the friction coefficient will continue at the level at which it was before. This observation is also shown in Figure 19.

The wear behavior of silicon nitride against silicon nitride is indicated in Table II. The areas of the wear scar on the pin were measured, larger areas indicating more wear. At this time the number of measurements is insufficient to provide appropriate confidence limits so that the numbers given are only indicative of trends. Thus the wear rate over the first minute of operation is much less under ethylene flow than without it at 600°C. At 500°C, the situation is reversed for the first minute, but is becoming nearly equal with and without ethylene after the first three minutes; for longer times the wear rate is again lower when ethylene is injected. Therefore ethylene

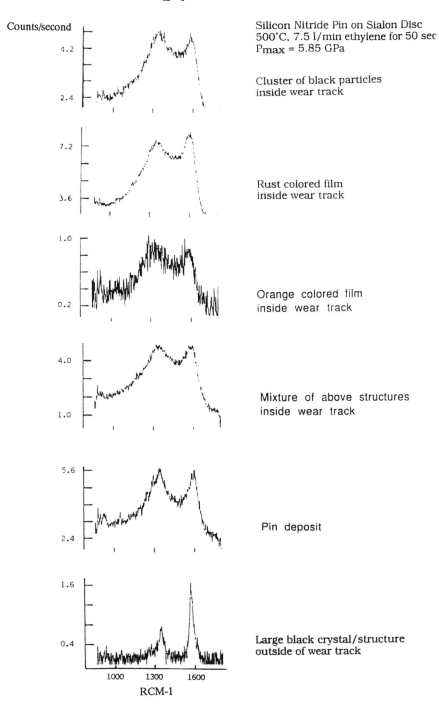

Counts/second

Silicon Nitride Pin on Sialon Disc
500°C, 7.5 1/min ethylene for 50 sec
P_{max} = 5.85 GPa

Cluster of black particles
inside wear track

Rust colored film
inside wear track

Orange colored film
inside wear track

Mixture of above structures
inside wear track

Pin deposit

Large black crystal/structure
outside of wear track

RCM-1

Figure 14.   Raman spectra of different carbons found in a pin-on-disc test of
Si3N4 on Sialon.

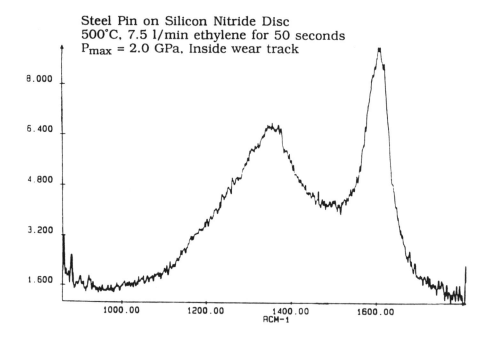

Steel Pin on Silicon Nitride Disc
500°C, 7.5 l/min ethylene for 50 seconds
P_{max} = 2.0 GPa, Inside wear track

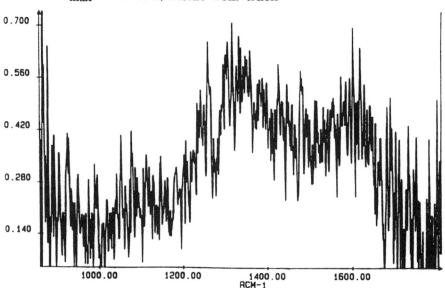

Steel Pin on Silicon Carbide Disc
500°C, 7.5 l/min ethylene for 50 seconds
P_{max} = 2.0 GPa, Inside wear track

Figure 15.   Raman spectra of carbon on a Si_3N_4 and a SiC disc in a wear track produced by a steel pin.

Counts/second

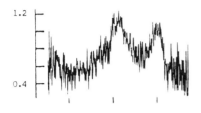

Steel Pin
(on Silicon Nitiride Disc)
500°C, 7.5 l/min ethylene for 50 secs
Pmax = 2.0 GPa
Area)pin scar = 0.6

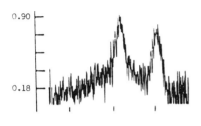

Sapphire Pin
(on Silicon Nitiride Disc)
500°C, 7.5 l/min ethylene for 50 secs
Pmax = 2.6 GPa
Area)pin scar = 0.72

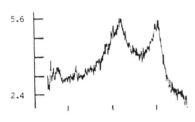

Silicon Nitride Pin
(on Sialon Disc)
500°C, 7.5 l/min ethylene for 50 secs
Pmax = 5.85 GPa
Area)pin scar = negligible

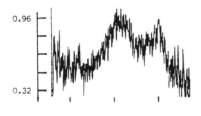

Sapphire Pin
(on Silicon Carbide Disc)
500°C, 7.5 l/min ethylene for 50 secs
Pmax = 4.4 GPa
Area)pin scar = 0.03

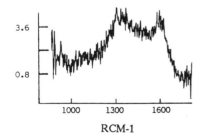

Silicon Nitiride Pin
(on Silicon Nitiride Disc)
500°C, 7.5 l/min ethylene for 50 secs
Pmax = 6.8 GPa
Area)pin scar = 0.54

RCM-1

Figure 16.   Raman spectra of different pins run with different plates. Note that the first (D) peak is always higher than the second (G) peak.

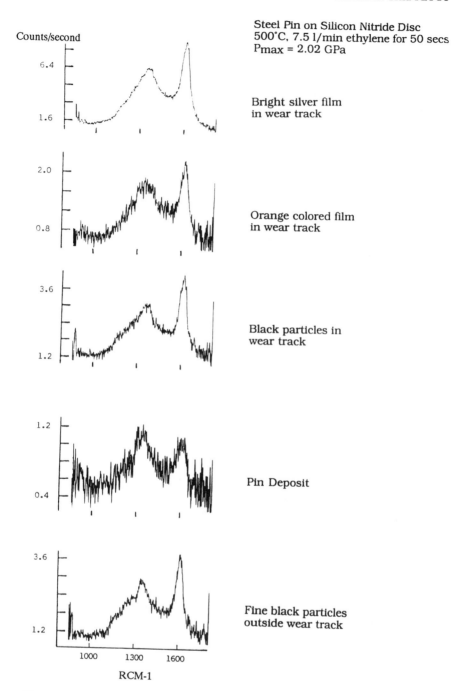

Figure 17.   Raman spectra of carbons on different carbons found in a pin-on-pin test of a steel disc on a Si3N4 plate.

Sapphire pin on Sialon disc
500°C, 7.5 l/min ethylene for 50 secs
Pmax = 2.62 GPa

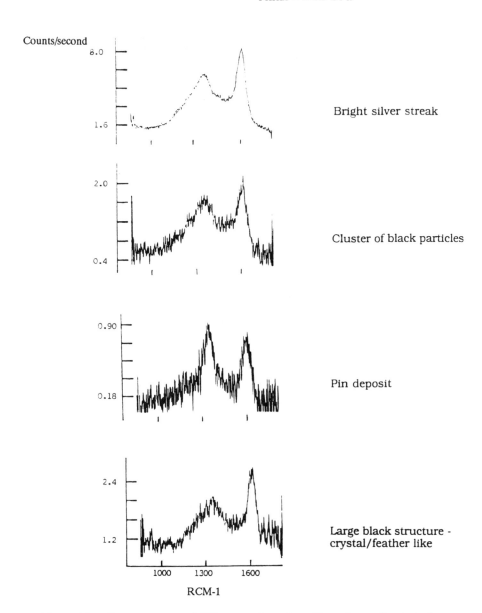

Counts/second

Bright silver streak

Cluster of black particles

Pin deposit

Large black structure - crystal/feather like

RCM-1

Figure 18.   Raman spectra of different carbons found in a pin-on-disc test of a sapphire pin on a Sialon disc.

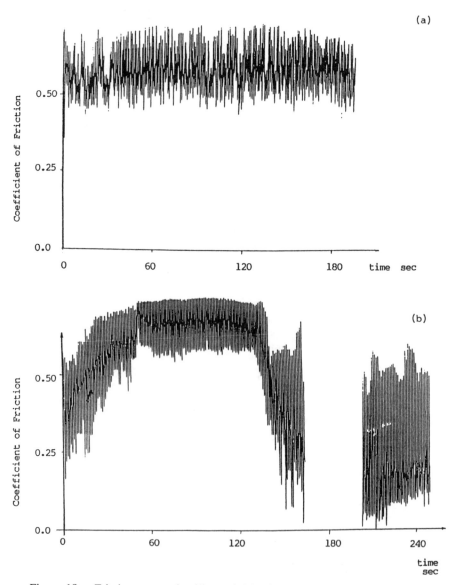

Figure 19.   Friction traces of a silicon nitride pin against a silicon nitride plate, (a) without ethylene flow, (b) with ethylene flow into conjunction region.

TABLE II
WEAR TEST RESULTS (preliminary)

(a) Silicon nitride pin against silicon nitride disc at 500°C under 7.5 l/min ethylene flow at different pressures

CONTACT PRESSURE	WEAR SCAR AREA (mm^2) Run Time			
	10 sec (69 RPM)	60 sec (69 RPM)	180 sec (69 RPM)	180 sec (129 RPM)
6.77 GPa	0.02	0.14		
2.83 GPa	0.02	0.14		
2.66 GPa	0.02	0.11	0.16	0.28
*2.66 GPa	0.02	0.04	0.12	0.31
*(no ethylene flow)				

NOTE: The wear scar under ethylene becomes smaller than that without ethylene with increasing wear time and speed, i.e. with longer track length.

(b) Silicon nitride pin against silicon nitride disc at 2.66 GPa and 69 RPM at different temperatures

CONTACT PRESSURE	Bulk Temp.	WEAR SCAR AREA (mm^2)		
		Run Time	(ethylene flow)	(no ethylene flow)
2.66 GPa	650°C	60 sec	0.02	0.05
	500°C	60 sec	0.11	0.04
	500°C	180 sec	0.16	0.12
	400°C	150 sec	0.12	0.11
	300°C	240 sec	0.06	0.14

NOTE: With ethylene flow, the wear scar area is a maximum at 500°C and lower both at temperatures below and above it. Without ethylene flow, the wear scar area increases with decreasing temperature. Ethylene flow reduces wear strongly at 650°C and above.

(c) Silicon nitride pin against both silicon nitride and silicon carbide discs at 500°C and at 2.66 GPa contact pressure and 69 RPM

MATERIALS	WEAR SCAR AREA (mm^2) Run Time	
	60 sec	180 sec
Si_3N_4 (ethylene flow)	0.11	0.16
Si_3N_4 (no ethylene flow)	0.04	0.12
SiC (ethylene flow)	0.12	0.19
SiC (no ethylene flow)	0.06	0.44

NOTE: The wear scar area is always less under ethylene flow than without ethylene flow under these.

flow reduces the wear rate at the highest pressures and temperatures and over long measurement times. Fortunately these are also the conditions required for practical use of the concept. More work will be needed to learn the effects of temperature, ethylene flow velocity and contact pressures [which also change with run time] on lubricating carbon formation and wear of silicon nitride on silicon nitride and other ceramic friction couples. Much of this work is already underway.

Preliminary results indicate that sapphire pins against both silicon nitride, Sialon or silicon carbide showed almost no wear under any of the conditions of Table II. The reason could be that sapphire is a single crystal while the other surfaces are polycrystalline.

DISCUSSION

Both metallic and ceramic surfaces can change structure under friction and wear already when unlubricated, especially at temperatures above 350°C. Raman and infrared spectroscopies can help with the identification of the changes, especially of crystal structure. In the presence of moisture and/or carbonaceous gases even more changes can be observed. Carbon formed from the gases can appear in many diverse forms, which are only now becoming appreciated. Many surface reactions occurring on both metals and ceramics have been extensively studied over the years from the point of view of catalysis, especially in petroleum refining, but not in relation to tribology. Carbon formation by catalysis has been studied only as a secondary issue. Carbon formed in the pyrolysis of gases had been considered to be independent of the nature of the bounding surfaces. A major exception has been the work of Tesner and his coworkers [28], who measured rates of carbon formation on various materials, such as cobalt and nickel, quartz, and carbon itself at temperatures above 700°C. They did not characterize the carbons formed, but they noted material dependences of formation rates.

Catalytic carbon formation on transition metal surfaces has been investigated for a long time but is still not fully understood. Just as is the case with pyrolytic carbon, the morphology of carbon deposited on nickel by exposure to hydrocarbons varies dramatically with deposition conditions [22]. The most dramatic finding of the many extensive studies has been the growth of carbon filaments with nickel crystallites attached to their tips. Sometimes these filaments evolve into clusters consisting of feather-like structures. We have observed such structures on nickel surfaces outside the wear track under some conditions. Filamentous carbon can be formed by diffusion of carbon through entrained small (~10 nm) nickel crystallites [21,22]. Raman spectra of filamentary carbons are consistent with the presence of very small graphitic crystallites (Figure 4).

Interestingly enough we have also found feather-like carbon structures formed from hydrocarbon gases on ceramics such as Si_3N_4, again only outside the wear track. Perhaps the element common to nickel and ceramics is the presence of carbides, which can be metal or silicon carbides. These carbides can be of varying stoichiometry and could therefore provide a high degree of adherence to the substrate. Surface carbon generated chemically is usually better adherent than carbon powder sprayed or even sputtered onto a surface.

We mentioned the possibility that the lubricating carbon formed at low temperatures, say around 200°C, was a friction polymer or a polymeric hydrocarbon surface film. Such a carbon deposit on nickel has sometimes been referred to in the literature as a beta-carbon. McCarthy et al [21,22] measured the amount adsorbed on the surface and found a mass of carbon less than expected for a bulk phase but more than expected for a surface layer and therefore identified it is polymeric. Our own investigations seem to agree with this assessment, but will need further study.

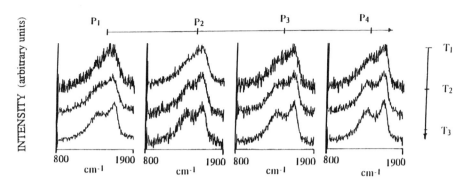

Figure 20.   Effect of substrate temperature ($T_1 < T_2 < T_3$) and argon pressure ($P_1 < P_2 < P_3 < P_4$) during carbon sputtering on the Raman spectra of carbon-overcoated magnetic storage discs. [Reproduced with permission from reference [18], Lauer, J.L. and DuPlessis, L., "Relation Between Deposition Parameters, Structure and Raman Spectra of Carbon Overcoats on Magnetic Storage Discs," *STLE, SP-29*, 71-78 (1990) (B. Bhushan, ed.).]

The carbon overcoats on hard computer discs, which are usually amorphous and very hard ("diamond-like") are generally sputtered onto an intermediate layer for better adherence. The wear characteristics of some of these carbons can be judged from their Raman spectra alone. Figure 20 shows typical Raman spectra of such carbon overcoats and Figure 21 identifies regions of high wear and low wear by both the frequency separation of the D and G peaks and the full width at half-height of the G band. Although a definitive analysis of these bands is not yet available, there is little doubt that both bands are really multiplets whose components vary in strength with number and orientation of the very small sp^2 and sp^3 crystallites in an amorphous matrix. Low wear would seem to imply low crystallinity and this would be reflected in the Raman spectrum.

Examination of Figure 21 shows that high wear is associated with large separation between the D and G peaks and with a wide D band. As mentioned, the D peak has been assigned to graphite edges or small graphite or diamond crystallites. The diamond band peaks at lower wavenumber (1332 cm⁻¹) than most D peaks (1340-1350 cm⁻¹). If diamond crystallites are in fact present along with graphite crystallites, they would have the effect of broadening the D band and shifting its peak toward lower wavenumbers, thereby increasing the D-G band separation. Higher diamond content would be consistent with high wear. However, this reasoning is still largely conjectural.

The line separating high and low wear regions in Figure 21 may appear somewhat arbitrary. However, if all the points representing the lowest wear were connected and the one point corresponding to FWHM = 380 cm⁻¹ were excluded, a very smooth curve would result. The line drawn took this point into account. Considering the difficulty of obtaining reproducible wear data, the scatter of points in Figure 21 is not excessive.

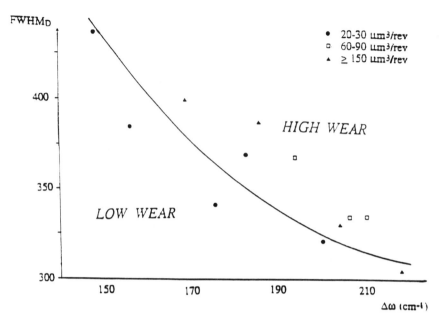

NOTE:

1. Wear rates above the indicated line are considered excessive.

2. Wear rates determined by optically measuring time for slider
    to penetrate Carbon coating, measuring volume, and assuming
    daily wear.

3. Density decreases with increasing band separation.

4. Wear increases with increasing density at constant band separation.

Figure 21.   Relation between the separation of the D and G Raman bands and
the full width at half maximum of the D band for different wear rates of the
magnetic storage discs of Figure 20.   [Reproduced with permission from
reference [18] Lauer, J.L. and DuPlessis, L., "Relation Between Deposition
Parameters, Structure and Raman Spectra of Carbon Overcoats on Magnetic
Storage Discs," *STLE, SP-29*, 71-78 (1990) (B. Bhushan, ed.).]

ACKNOWLEDGEMENT

This work was partially funded by the Rensselaer Polytechnic Institute High
Temperature Technology Program, administered by the New York State Energy
Research and Development Authority (NYSERDA) and by the National Science
Foundation under Grant No. MSM-87-1281. Any opinions, findings and conclusions
or recommendations expressed in this publication are those of the authors and do not
necessarily reflect the views of the National Science Foundation.

REFERENCES

1. Lauer, J.L., and Bunting, B.G., "High Temperature Solid Lubrication by Catalytically Generated Carbon," *Tribology Transactions 31,* 338-349 (1988).
2. Graham, E.E. and Klaus, E.E., "Lubrication from the Vapor Phase at High Temperatures," *ASLE Transactions 29*(2), 229-234 (1986).
3. Lauer, J.L., and Dwyer, S.R., "Tribochemical Lubrication of Ceramics by Carbonaceous Vapors," STLE Preprint No. 90-TC-5A-1 (Toronto, October 1990).
4. Kinney, C.R., "Studies on Producing Graphitizable Carbons," *Proceedings of the First and Second Conferences on Carbon,*" The University of Buffalo, 1956. Printed by the Waverley Press, Inc., Baltimore, pp. 83-92.
5. Tsai, H-C, Bogy, D.B., "Critical Review: Characterization of Diamondlike Carbon Films and Their Application as Overcoats on Thin-Film Media for Magnetic Recording," *J. Vac. Sci. Technol.* A5(6), 3287-3312 (1987).
6. Eversole, W.G., U.S.P. #3,030,187 and #3,030,188 (1962). [Assigned to Union Carbide Corporation].
7. Lauer, J.L., "Preparation of Crystalline Carbonaceous Materials," U.S.P. #3,362,788 (1968). [Assigned to Sun Oil Company].
8. Lauer, J.L. and Jones, W.R. Jr., "Friction Polymers," ASLE Special Publication No. 21, Tribology of Magnetic Recording Media, p. 14-23 (1986).
9. Furey, M.J., "The Formation of Polymeric Wear Films Directly on Rubbing Surfaces to Reduce Wear," *Wear 26,* 369-392 (1973).
10. Hegemann, B.E., Jahanmir, S., and Hsu, S.M., "Microspectroscopy Applications in Tribology," *Microbeam Analysis 23,* 193-195 (1988).
11. Exarhos, G.and Donley, M.S., "Real-Time Raman Detection of Molecular Changes in Ceramics," *Microbeam Analysis 22,* 125-127 (1987) (R.H. Geiss, ed).
12. Kinoshita, K., *Carbon Electrochemical and Physicochemical Properties,* John Wiley & Sons, New York, NY 1988.
13. Knight, D.S., and White, W.B., "Characterization of Diamond Films by Raman Spectroscopy," *J. Mater. Res. 4*(2), 385-393 (1989)
14. Strel'nitskii, V.E., et al., "Study of Infrared Emission Spectra of Carbon Condensates," *Sverkhtverd. Mater.,* 1986 (6), 7-12; *CA 107*: 4849t.
15. Lespade, P. and Al-Jishi, R., "Model for Light Scattering from Incompletely Graphitized Carbons," *Carbon 20*(5), 427-431 (1982).
16. McCreery, R., Wang, Y., and Boling, R., "Raman Spectroscopy of Carbon Materials: Frequency Shifts and Resonance Effects over a 351 nm - 1064 nm Laser Wavelength Range," Paper No. 651, Abstracts of the 1990 Pittsburgh Conference on Analytical Chemistry and Applied Spectroscopy (March 5-9, 1990).
17. Seki, H., "Raman Spectroscopy of Carbon Overcoats for Magnatic Disks," *Surface and Coatings Technology, 37,* 161-178 (1989).
18. Lauer, J.L. and DuPlessis, L., "Relation Between Deposition Parameters, Structure and Raman Spectra of Carbon Overcoats on Magnetic Storage Discs," *STLE, SP-29,* 71-78 (1990) (B. Bhushan, ed).
19. Grill, A., Patel V., and Meyerson, B.S., "Optical and Tribological Properties of Heat-Treated Diamond-like Carbon," *J. Mater. Res., 5*(11), 2531-2537 (1990).
20. Krishnan, K., and Hill, S.L., "FTIR Microsampling Techniques." Chapter 3 in *Practical Fourier Transform Spectroscopy.* Academic Press, New York, NY 1990, pp. 103-165 (See in particular p. 121, Fig. 11).
21. McCarthy, J.G., and Wise, H., *J. Catal. 57,* 406 (1979)

22. McCarthy, J.G., Hou, P.Y., Sheridan, D., and Wise, H., "Coke Formation on Metal Surfaces," *ACS Symposium Series 202*, 253-282.
23. Lacava, A.J., et al, *ACS Symposium Series 202*, 89-107 (1982).
24. Lauer, J.L. and Dwyer, S.R., "Continuous High Temperature Lubrication of Ceramics by Carbon Generated Catalytically from Hydrocarbon Gases," *Tribology Transactions 33*(4), 529-534 (1990).
25. Lauer, J.L., Final Report on Army Research Office, Grant #DAAL03-86-K-076, Jan. 14, 1991
26. Cotton, F.A. and Wilkinson, G., *Advanced Inorganic Chemistry*, Fifth Edition, John Wiley & Sons, New York, 1988, p. 241.
27. Derbyshire, F.J., Presland, A., and Trimer, D.L., *Carbon 13,* 11 (1978).
28. Tesner, P.A., "Kinetics of Pyrolytic Carbon Formation," in *Chemistry and Physics of Carbon* (Peter A Thrower, Ed.) Marcel Dekker Publ., New York, NY 1988., Vol. 19, p. 65-161.

RECEIVED October 21, 1991

LUBRICANT–SURFACE INTERACTIONS

Chapter 8

Highly Chlorinated Methanes and Ethanes on Ferrous Surfaces

Surface Decomposition and Extreme-Pressure Tribological Properties

P. V. Kotvis[1], L. Huezo, W. S. Millman, and W. T. Tysoe[2]

Department of Chemistry and Laboratory for Surface Studies, University of Wisconsin, Milwaukee, WI 53211

Chlorinated hydrocarbons are an important class of extreme pressure (EP) additives to metalworking fluids for preventing seizure between the contacting metal surfaces. It is proposed that they function by thermally decomposing at the interface between these surfaces, forming an iron halide + carbon anti-seizure film. This film is continually removed by the frictional forces acting at the surface and the net film thickness is the result of a dynamic balance between growth and removal. The film growth and removal rates are measured independently, and the film thickness is calculated from these data as a function of applied load. The experimental data for seizure load versus additive concentration is well reproduced by assuming that seizure occurs when the thickness of this protective film diminishes to zero.

Lubricants are very often used by metalworking industries for operations such as machining, wire drawing etc. to increase production (by increasing machining speeds, reducing down-time) and to improve the quality of the final product. The market for such lubricants exceeds $200 million annually in the U.S. alone. The base fluid or major component of these lubricants is usually a petroleum vacuum distillate (often referred to as mineral oil) or water. Certain chemicals added to the base fluid provide other desirable features, in particular, prevention of excess wear or even seizure of the contacting surfaces under extreme pressure. These "extreme pressure" additives often consist of chlorinated hydrocarbons. However, their role in mediating lubrication under conditions of extreme pressure (EP) is

[1]Current address: Frank P. Farrell Laboratory, Benz Oil, Inc., 2724 West Hampton Avenue, Milwaukee, WI 53209
[2]Corresponding author

not well understood. In addition, extreme pressure lubrication is ideally suited for fundamental study since the extremely high applied load results in significant material removal from the surfaces so that the interfacial chemistry is likely to reflect that of the clean metal.

Synthesis and Characterization of a Model Extreme Pressure Lubricant

A model extreme pressure lubricant that mimics the properties of commercial lubricants rather well has been synthesized and consists of a well-defined chlorinated hydrocarbon dissolved in a poly-α-olefin (PAO) (*1*). The olefin is 1,3,5-tri-n-octyl hexane ($C_{30}H_{62}$), which was selected since it has properties (i.e., viscosity, density etc.) close to those of a commercial mineral oil, and can be obtained with high purity (>99.5%). Its extreme pressure lubricating properties were measured using a pin and v-block apparatus (shown in Figure 1). Here the pin and v-blocks are immersed in the model EP lubricant where the pin rotates at 290 revolutions per minute and the load applied to the v-blocks can be varied. The fluid is held constant to within ± 1K using a recirculating temperature controller. Care is taken to completely exclude oxygen and water. In order to obtain reproducible results, the apparatus was initially run for 300 s using a force of 120 kg. The jaw load is then increased linearly so that the torque required to maintain the rotational motion of the pin also increases. At a critical load the torque suddenly increases, indicating that the lubricant has failed and that the system has "seized". This "seizure load" L_s is taken as a measure of the effectiveness of the extreme pressure lubricant. A higher seizure load implies a more effective lubricant additive. Shown in Figure 2 are plots of seizure load versus additive concentration obtained using a range of chlorinated hydrocarbon additives. Clearly in all cases, the seizure load is higher when the additive is present indicating that chlorinated hydrocarbons are indeed effective anti-seizure (EP) additives. Note that the abscissa has been normalized to indicate the same chlorine content for each additive.

The curves for methylene chloride and chloroform show an initial sharp increase in seizure load with additive concentration and remain rather constant thereafter. This behavior is designated Type I. In contrast to this, when CCl_4 is used as an additive, *no* plateau is observed and the seizure load continually increases with concentration up to the maximum load that can be applied using the pin and v-block apparatus. This is designated Type II behavior, and indicates that CCl_4 is an outstandingly effective extreme pressure additive. This result is in agreement with other work (*2*). Also shown in Figure 2 are curves for additives that display both types of behavior (e.g., C_2Cl_6). In this case, at lower concentrations the seizure load initially increases and then forms a plateau. At higher concentrations, however, the seizure load starts to increase once again, in a manner akin to a Type II additive (CCl_4).

Analysis of the surfaces using both Auger and X-ray photoelectron spectroscopy (XPS) reveals the presence of carbon and chlorine. The resulting Cl 2p spectra are shown in Figure 3 compared with spectra for $FeCl_2$, $FeCl_3$ and polyvinylchloride. These results suggest that the chlorine is present as a halide.

Interfacial Temperature Measurements

The temperature at the interface between the pin and the v-block increases as a function of applied load because of the increased frictional energy that is dissipated. The temperature *adjacent* to the interface was measured by means of a chromel/alumel thermocouple spot-welded to the stationary v-block and as close as possible to the pin. While the temperature measured in this fashion will significantly underestimate the actual interfacial temperature it will effectively indicate changes in the temperature at the interface. Figure 4 shows a plot of the thermocouple temperature measured as a function of applied load when using 0.75% of 1,4-dichlorobutane as additive. This corresponds to Type I behavior in the plateau region. In this case the temperature increases linearly with applied load until seizure when the temperature increases drastically. The thermocouple temperature is ⁻390 K at this point. In the case of Type II behavior (C_2HCl_5; 2.4%), the temperature increases linearly at lower loads and shows a similar rapid temperature increase when the thermocouple reads 390 K but which, in this case, does not lead to seizure. Rather the temperature stabilizes and continues its linear increase at higher applied loads. These results suggest that, at least in the case of a Type I additive in the plateau region, seizure might occur when the interface reaches a critical temperature, possibly corresponding to some physico-chemical change at the surface. The results shown in Figure 4 suggest that a similar change takes place when a Type II additive is used which, in that case, does not lead to seizure. If so, increasing the fluid temperature should *decrease* the seizure load since the critical temperature should be attained at a lower applied load. This effect is illustrated in Figure 5, which displays a plot of seizure load versus bath temperature when using 3.0 wt % of 1,4-dichlorobutane (corresponding to a Type I additive in its plateau region). Clearly the seizure load does decrease with increasing bath temperature, confirming that seizure does occur when the interface attains a critical temperature. The power dissipated at the interface will increase linearly with applied load L for a constant coefficient of friction, so that the interfacial temperature T can be written as:

$$T = T_0 + aL \tag{1}$$

where a is a constant which depends on the experimental geometry and T_0 is the bath temperature. If seizure occurs when the interface reaches a critical temperature T_c, then the corresponding seizure load L_s is given by:

$$T_c = T_0 + aL_s \tag{2}$$

and implies that the seizure load will vary as the bath temperature according to:

$$L_s = -T_0/a + T_c/a \tag{3}$$

in accord with the data shown in Figure 5. Measurement of the slope and intercept of a linear fit to the results of Figure 5 indicate that a = 2.5±0.4 K/kg and $T_c = 950 \pm 100$ K. Thus the system seizes when the interface reaches ⁻950 K. This will be discussed in greater detail below.

Surface Chemistry of Chlorinated Hydrocarbons

It is argued above that the chemistry at the interface between the pin and the v-block reflects that of the clean surface since material is continually removed. Analysis of the pin or v-block surfaces (Figure 3) indicates the presence of carbon and an iron halide, and the results of the previous section suggest that the interface can reach temperatures of several hundred degrees during the tribological experiment. It is likely, therefore, that the chlorinated hydrocarbon additive can thermally decompose at the pin and v-block interface forming a halide + carbon layer which acts to prevent seizure. The thermal decomposition kinetics of a range of chlorinated hydrocarbons on a clean iron foil were therefore measured using a microbalance. In this case, film growth kinetics can be continually monitored from mass changes in the sample and the film thickness can be estimated by assuming that it consists of $FeCl_2$. This is likely to result in some error in measurement of absolute thicknesses (since the film undoubtedly incorporates carbon). Relative film thicknesses (and growth rates) will nevertheless be reliable. Figure 6 shows plots of film thickness (calculated from the sample mass change, normalized to reactant pressure) as a function of time for the thermal decomposition of several volatile chlorinated hydrocarbons on an iron foil heated to 746 K. CCl_4 is by far the most reactive chlorinated hydrocarbon. Both chloroform and methylene chloride are significantly less reactive than carbon tetrachloride, and films are formed slightly more rapidly from the thermal decomposition of $CHCl_3$ than CH_2Cl_2. Note that the relative reactivities of each of these chlorinated hydrocarbons for film growth correspond well to their relative efficacies as extreme pressure additives (Figure 2) implying that film growth is important in extreme pressure lubrication. Analysis of the surfaces of the foils using XPS reveals the presence of both carbon and an iron halide at the surface in accord with the composition of the surfaces of the pins and v-blocks (Figure 3) further confirming that the film growth chemistry is identical in both cases.

The film growth kinetics were measured for CH_2Cl_2, a Type I additive, as a function of both sample temperature between 400 and 700 K and reactant pressure. A film growth activation energy of 8.1 ± 0.4 kcal/mol was measured for the reaction. The growth rate pressure dependence was monitored by measuring the initial film growth rate at a sample temperature of 620 K and shows that the film growth reaction is first order in reactant pressure (or concentration). The initial growth kinetics can be summarized as:

$$\text{Rate} = A \, [CH_2Cl_2] \exp(-E_{act}/RT) \qquad (4)$$

where E_{act} is the growth activation energy (8.1 ± 0.4 kcal/mol), $[CH_2Cl_2]$ is the methylene chloride concentration and A the rate pre-exponential factor. Note that this factor depends on the accuracy of the film thickness measurements

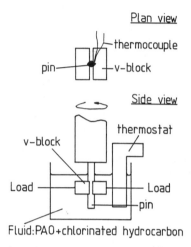

Figure 1.   Schematic diagram showing the pin and v-block apparatus. (Reproduced by permission from ref. 3. Copyright 1991, Elsevier Science Publishing.)

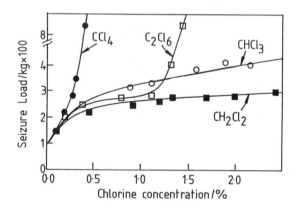

Figure 2.   Plot of seizure load versus additive concentration for various chlorinated hydrocarbon additives. (Reproduced by permission from ref. 3. Copyright 1991, Elsevier Science Publishing.)

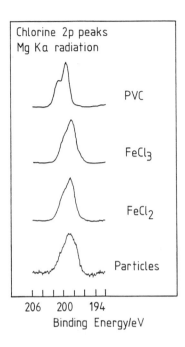

Figure 3. Chlorine 2p x-ray photoelectron spectrum of the wear particle surface compared with the corresponding spectra for $FeCl_2$, $FeCl_3$ and polyvinylchloride. (Reproduced by permission from ref. 1. Copyright 1989, Elsevier Science Publishers (North Holland))

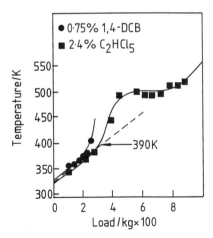

Figure 4. Plot of the temperature close to the interface between the pin and v-block as a function of applied load using 0.75 wt.% 1,4-DCB (Type I) and 2.4 wt.% C_2HCl_5 (Type II) as additives. (Reproduced by permission from ref. 3. Copyright 1991, Elsevier Science Publishing.)

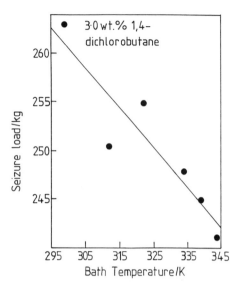

Figure 5. Plot of seizure load versus fluid temperature obtained using 3.0 wt.% of 1,4-DCB as additive. (Reproduced by permission from ref. 3. Copyright 1991, Elsevier Science Publishing.)

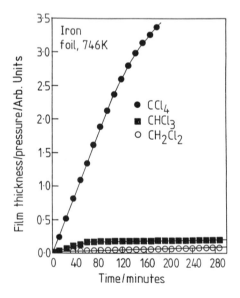

Figure 6. Film thickness (measured from the mass change, see text) plotted versus time for the thermal decomposition of several chlorinated hydrocarbons on an iron foil. (Reproduced by permission from ref. 3. Copyright 1991, Elsevier Science Publishing.)

which depends on the assumption that the film consists entirely of an iron halide. These results, however, clearly demonstrate that chlorinated hydrocarbons thermally decompose on an iron surface over the temperature range encountered in extreme pressure tribology to deposit an iron halide + carbon layer. These results will be compared in a more quantitative fashion with the tribological results below.

Film Removal During Extreme Pressure Lubrication

In addition to film formation during extreme pressure lubrication by the thermal decomposition of the chlorinated hydrocarbon additive, material is also removed from the surface of both the pin and the v-blocks (*3*). Indeed, debris is detected at the bottom of the container following an experiment using the pin and v-block apparatus. The particle size distribution can be measured using scanning electron micrographs of the wear debris and shows an average value of ~5.1μm for the size of the particles removed using only PAO. 5% of 1,4-dichlorobutane added to the fluid considerably alters the shape of the distribution and yields an average particle size of 2.6μm in the debris. Therefore the presence of an EP additive decreases the size of the particles removed during lubrication.

In addition, the width of the wear scar on the surface of the v-block can be used to measure the rate of material removal. The volume V of material removed from the v-block can be calculated from the width of the wear scar w, and the volume of material removed in 600 s when using 2.5 wt.% of CH_2Cl_2 as additive is displayed in Figure 7. The rate of material removal clearly increases rapidly as a function of applied load. The volume of material V of shear strength S removed between two sliding surfaces per unit sliding length with an applied load L is usually described by (*4,5*):

$$V = kL/S \tag{5}$$

where k is a constant. In the case of the pin and v-block experiments, the angular velocity of the pin is held constant so that, for a constant pin diameter, the tangential velocity is constant and equation 5 can be rewritten as:

$$dV/dt = c\,L/S \tag{6}$$

where c is a constant that depends on k, the pin diameter and angular velocity. However, the temperature at the interface also increases with applied load L (according to equation 1). The shear strength S of the interfacial material is temperature dependent (decreasing to essentially zero as the material melts) so that S also depends on L. Thermodynamic arguments have been used to demonstrate that shear strength varies with temperature T as:

$$S = S_0\,\ln(T_m/T) \tag{7}$$

where T_m is the melting point of the interfacial material (*6*). The overall dependence of the removal rate on applied load can then be obtained by substituting equations 1 and 7 into 6 to yield:

$$dV/dt = K \, L/ln[T_m/(T_0 + aL)] \qquad (8)$$

where K is a constant. The best fit function of this type is shown plotted onto the data of Figure 7 and the agreement between theory and experiment is very good. The parameters used to obtain this fit were T_0 = 300 ± 10 K, a = 2.4 ± 0.2 K/kg, T_m = 916 ± 10 K and K = 1.32 ± 0.05x10^{-5} mm^3/kg/min. The value of a is in good agreement with that determined above from the variation in seizure load with bath temperature (Figure 5) and the value of T_0 is good agreement with the bath temperature which was held at 300 K. More important, the value for the interfacial melting point (T_m ⁻ 916 K) is in excellent agreement with the critical seizure temperature measured above (T_c = 950 K). These results then suggest that the critical temperature is associated with the onset of melting of the interfacial material. Note that this temperature is in good agreement with the melting point of FeCl$_2$ (943 K; (7)) suggesting that this forms the tribologically significant layer, in accord with surface analyses of the pins and v-blocks (Figure 3) and the nature of the film formed by the thermal decomposition of chlorinated hydrocarbons.

Model for the Activity of Methylene Chloride as an Extreme Pressure Additive

The picture that emerges for the activity of a chlorinated hydrocarbon as an extreme pressure lubricant additive is that of a dynamic process where a chlorinated hydrocarbon additive thermally decomposes at the interface between the pin and the v-block to form a carbon- and iron chloride-containing film. This appears to produce a tribologically active layer. This film is continually removed by the relative motion of the pins and v-blocks so that the resultant film thickness X at the interface can be viewed as an equilibrium between its rate of formation and removal. If the film growth rate is r_g and the removal rate is r_r, the film thickness at any time is given by:

$$dX/dt = r_g - r_r \qquad (9)$$

where the functional forms of r_r and r_g have been established as described above using the pin and v-block apparatus and microbalance experiments respectively.

The observation that, in the case of a Type I additive in the plateau region, seizure occurs when the iron chloride film melts so that the protective film is removed from the surface, suggests that seizure occurs when the film thickness X diminishes to zero. Equation 9 can then be solved and the load at which X is zero is taken to be L_s the seizure load. It should be emphasized that all the parameters required for equation 9 have been independently established. Note that a run-in period is required in the pin and v-block experiments in order to obtain reliable data. It is assumed that an initial film of thickness X_0 is formed during this period and that an initial wear scar of width w_0 is also formed. The value of w_0 is calculated from equation 10, and X_0 is allowed to vary to yield the best agreement between the experiment and the simulation. The value of A, the pre-exponential factor in the growth rate equation (equation 4), is allowed to vary around the experimentally determined value for the thermal decomposition of methylene chloride vapor for two reasons: (i) the value has been calculated assuming that a pure $FeCl_2$ layer is deposited whereas the layer contains carbon, and (ii) the absolute decomposition rate is likely to be modified by solution in the base fluid. All other parameters are those determined above so that equation 9 is solved numerically using only two adjustable parameters.

Figure 8 shows the results of the simulation compared with the experimental results for the seizure load as a function of methylene chloride concentration. The agreement is extremely good and suggests that the assumptions used in the simulation are valid. The best fit between the simulation and the experimental data is obtained using a value for X_0 of $0.22 \mu m$, suggesting that a film of approximately this thickness is deposited during the "run-in" period. Figure 9 shows corresponding plots of film thickness for various additive concentrations.

Conclusions

A model is proposed for the operation of methylene chloride (CH_2Cl_2) as an extreme pressure additive in which a tribologically active film consisting of an iron halide plus carbon is deposited at the interface by the thermal decomposition of the chlorinated hydrocarbon additive. The effect of this film is to protect the metal surfaces from seizure and is it continually removed during the experiment. The net film thickness is given by a dynamic balance between the rate of film formation by additive decomposition and film removal by friction. The tribological behavior of CH_2Cl_2 as an additive can be effectively simulated using these assumptions to yield the experimentally observed plot of seizure load versus additive concentration. Other additives which have plots of seizure load versus additive concentrations that are very similar to the behavior for methylene chloride are 1,4- and 2,3-dichlorobutane (both isomers display essentially identical behavior), 1-chlorodecane and chloroform. Note, however, that the plateau region for chloroform is slightly higher than the others.

The interfacial temperature has been measured independently using two different methods (i.e., from the variation of seizure load with bath temperature and from the film removal kinetics), both of which indicate that the interfacial temperature varies linearly as a function of applied load such that an interfacial temperature of ~950 K corresponds to an applied load of 260 kg, the seizure load in the plateau region when methylene chloride is used as additive. This temperature (and applied load) corresponds to the melting point of $FeCl_2$, strongly implying that this is the tribologically significant film. Note, however, that both in the case of Type II additives and some Type I additives (notably $CHCl_3$) the applied load can exceed 260 kg without seizure. Indeed, when CCl_4 is used as an additive above 0.5 wt.% chlorine concentrations, the seizure load exceeds 800 kg. According to equation 1 the interfacial temperature is ~2300 K

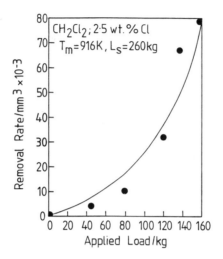

Figure 7. Plot showing the film removal rate as a function of applied load using 2.5 wt.% CH_2Cl_2 as additive.

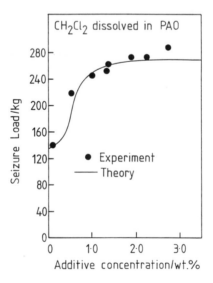

Figure 8. Comparison of a simulation (solid line) of seizure load versus additive concentration with experimental data for CH_2Cl_2 (•).

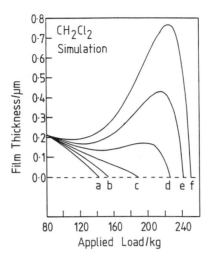

Figure 9. Plot of the film thickness versus applied load for various CH_2Cl_2 concentrations (0.2 % apart, starting at 0.2 wt.%) obtained with the parameters used for the simulation shown in Figure 8.

at this point. This is far in excess of the melting point of iron chloride. These results strongly imply that some other tribologically significant layer is formed under these conditions and that the melting point of this layer exceeds 2300 K. Possible candidates for this layer are carbon or iron carbide, both of which have extremely high melting points.

Literature Cited

1. Kotvis, P.V.; Tysoe, W.T. *Appl. Surf. Sci.* **1989,** *40,* 213 and references contained therein.
2. Shaw, M.C. *Ann. N.Y. Acad. Sci.* **1951,** *53,* 962
3. Kotvis, P.V.; Huezo, L.; Millman, W.S.; Tysoe, W.T. *Wear* in press
4. Rabinowitz, E. *Friction and Wear of Materials;* John Wiley and Sons: New York, NY, 1965
5. Holm, R. *Electrical Contacts;* Almquist and Wiksells: Stockholm, 1946
6. Ernst, H.; Merchant, M.E. *Proc. Special Summer Conference on Friction and Surface Finish, M.I.T.* **1940,** 76
7. *CRC Handbook of Physics and Chemistry;* Waest, R.C.; Selby, S.M., Eds.; CRC Press: Cleveland, OH, 1968

RECEIVED October 21, 1991

Chapter 9

Ultrathin Perfluoropolyether Films

J. Rühe[1,3], S. Kuan[2], G. Blackman[1], V. Novotny[1], T. Clarke[1], and G. Bryan Street[1]

[1]IBM Almaden Research Center, 650 Harry Road, San Jose, CA 95120
[2]IBM Storage Systems Products Division, 5600 Cottle Road, San Jose, CA 95124

Perfluoropolyethers with functional endgroups can be attached to various metallic and non–metallic surfaces by thermal treatment. The surface chemistry of the substrate has a strong influence on the amount of polymer attached and on the kinetics of the surface reaction. The perfluoropolyether films consist of chemisorbed as well as physisorbed polymer. Consequently the strength of attachment of the different polymer molecules varies widely. Physisorbed polymer is mobile and can be displaced by polar low molecular weight compounds such as water and alcohols. The attached polymer films were investigated by FT–IR, ellipsometry, microcalorimetry, XPS and atomic force microscopy.

Films with thicknesses of the order of monolayers attached to the surface of overcoats of thin film magnetic disks can enhance the life time of the magnetic medium dramatically (1–3). Such films can consist of low molecular weight compounds such as stearic acid (3), self–assembled layers of alkylsilanes (4), Langmuir–Blogett–films of salts of fatty acids (5–6) or polymers, mostly perfluoropolyethers (7–10). To avoid depletion of the polymers due to spin–off or displacement through contaminants adsorbing from the ambient, the films have to be strongly attached to the surface. It has been demonstrated that perfluoropolyethers with appropriate functional endgroups such as hydroxy–, carboxylic acid– or piperonyl groups can be firmly attached to the surface of carbon overcoats of magnetic media by thermal treatment (7–8).

[3]Permanent address: Universität Bayreuth, Makromol. Chemie II, Postfach 101251, 8580 Bayreuth, Germany

Considerable efforts have been made to elucidate the tribological properties of thin films attached to the surface of carbon overcoats. However knowledge about the attachment of the polymer to the surface is surprisingly sparse and little is known about the surface chemistry of the attachment reaction. Polymer which cannot be washed off during a standardized cleaning procedure after thermal treatment is usually assumed to be covalently bonded. This is not necessarily the case as chemisorbed and strongly physisorbed polymer is difficult to distinguish if no good solvent for the polymer can be found. In fact in the samples we examined the attached films consisted of chemisorbed and physisorbed polymer. The strength of the adsorption depends on the chain geometry of the polymer at the surface (especially the lengths of the trains and loops) and subsequently a different displacement and dissolution behavior can be observed.

Furthermore little attention has been paid to the influence of the surface chemistry of the overcoat. Frequently the overcoat is employed without any further characterization of the surface groups. We found however, that the surface chemistry has a very pronounced influence on the attachment of the polymer and its displacement from the surface. Therefore samples with an identical history (thermal treatment, solvent exposure), but different numbers of reactive groups at the surface show very pronounced differences in the strength of the attachment of the polymer to the surface.

In this work we describe studies on adsorption and chemisorption of various perfluoropolyethers on carbon and other non–metallic and metallic surfaces using microcalorimetry, ellipsometry and FT–IR. In addition, the obtained ultrathin polymer films were investigated using XPS, ellipsometry, FT–IR and atomic force microscopy.

Sample Preparation

The polymers employed in this study are depicted in table 1. They had similar polymer backbone structures and comparable molecular weights but different endgroups (hydroxyl– vs. trifluoromethyl–endgroups). Depending on the desired initial film thickness the polymers were coated on the surface of the substrate by wiping (for thick polymer films >1 μm) or dip coating (typical film thickness 10 nm). In dipping experiments concentration (0.5 – 5 g/l of the perfluoropolyether in Freon) and withdrawal speed (1 – 3 mm/s) were adjusted in order to obtain the desired thickness. After thermal treatment the samples were washed carefully and dried in a stream of filtered, dry nitrogen.

The surfaces were cleaned prior to deposition of the polymer from organic contaminants, which adsorb rapidly from the ambient. Contaminants on carbon surfaces were removed using either deep–UV–irradiation (wavelength 185 nm) under nitrogen (to avoid generation of ozone) or by washing with detergent and deionized water. Samples were rinsed thoroughly prior to use with chloroform and trichlorotrifluorethane. Metal surfaces were cleaned with UV and ozone. The resulting surfaces were highly hydrophilic with wetting angles < 5⁰. Carbon films (about 25.0 nm thick) were sputtered on Si–wafers (for ellipsometry measurements) or highly reflecting aluminum–discs (for FT–IR measurements).

Table 1: structures, trade names and molecular weight of the polymers used

structure	trade name	molecular weight (M_w)
$HO-CH_2-CF_2 - O(-CF_2-O-)_m$ $HO-CH_2-CF_2 - O(-CF_2-CF_2-O-)_n$ $\underline{1}$	ZDOL	2200 4000
$CF_3-O(-CF_2-CF_2-O-)_n(-CF_2-O-)_m CF_3$ $\underline{2}$	Z 03 Z 15	3700 9300

$m,n = 0.4–0.6$

Attachment of perfluoropolyether films to different surfaces

Polymer films of $\underline{1}$ (ZDOL) were attached to various metallic, non–metallic and polymer surfaces. Table 2 depicts the resulting film thicknesses on the various substrates. All samples were prepared by dip–coating of a thin film of the polymer on the substrate, heating for one hour to 150 °C in air and extracting non–attached polymer by washing with solvent.
It should be noted that the thicknesses of the attached layers were strongly dependent on the surface chemistry of the substrate. For example the thicknesses of the PFPE–films attached to carbon varied by ± 25% when samples prepared under different sputtering conditions and subsequently handled under identical conditions.

The numbers of hydroxyl, carbonyl– and carboxylic acid groups on the carbon surfaces were determined by XPS–analysis after surface derivatization with fluorinated tag molecules. It was found that all surfaces were highly functionalized, but the exact number of the corresponding groups varied from batch to batch and was strongly dependent on the conditions of preparation of the sample. A detailed description of the experiments is reported elsewhere (11).

When the reactive groups on the surface were blocked (for example through reaction of the hydroxylgroups on SiO_2 surfaces with hexamethyldisilazane) or only very few reactive surface groups were available (as in polystyrene for example) very little polymer (< 0.5 nm) was found on the surface after thermal treatment and solvent extraction.

Table 2:   Thickness of the perfluoropolyether films attached to various
surfaces and activation energy for the attachment

material	thickness (nm)	method of measurement	activation energy[*]
carbon (sputtered)	1.8	FT–IR,XPS,ellips.	5±2
graphite	1.5	XPS	–
silica	2.5	ellips.,XPS	15±2
aluminum	4.0	FT–IR	16±4
gold	1.7	ellips.	–
chromium	4.5	FT–IR	14±5
poly(acrylic acid)	2.6	XPS	–
poly(vinyl alcohol)	1.8	XPS	–
polystyrene	<.5	XPS	–

[*] activation energy in kcal/mol

To obtain the activation energy of the surface reaction on carbon, SiO_2 and aluminum the surface reaction was quenched after different lengths of time at varying temperatures. The results of such measurements are depicted in fig. 1, which shows the thickness of PFPE layers attached to a silica surface as a function of reaction time and temperature. Additionally the reaction rate at the beginning of the reaction is shown. The activation energy was obtained through an Arrhenius–plot of the reaction rates at different temperatures. The value of the activation energy for the reaction of 1 with a silica surface (15 kcal/mol) is in good agreement with the activation energy of a surface–esterification of silica with low–molecular weight alcohols (10–20 kcal/mol) (12). A detailed discussion of the kinetics of the polymer attachment will be published elsewhere (13).

During the thermal treatment procedure used to attach the polymer 1 (ZDOL), the molecular weight of the PFPE increases significantly through evaporation of lower molecular weight fractions. End–group analysis of extracted polymer with [1])F–NMR shows that the average molecular weight increases upon heating at 150 °C for 1 hour from approximately 2200 g/mol to about 6000 g/mol. The sample thickness in these experiments was 10 nm of 1 (ZDOL).

Furthermore it could be seen from ^{19}F–NMR–experiments that under the conditions employed (150 °C, 1 hr.) no degradation of the PFPE occured.

However it should be noted, that at higher temperatures (>220 ^{0}C) or in the presence of very strong acids such as H_2SO_4 or Lewis–acids such as $TiCl_4$ degradation does occur and the molecular weight of the polymer decreases significantly.

Physisorbed Polymer

It is important to note that not all of the polymer, which cannot be extracted from the attached film by a short solvent wash, is chemisorbed on the surface. This can be seen quite clearly if the polymer films are extracted with solvents for a prolonged time or are exposed to low molecular weight displacers. It was found, that extended solvent exposure (Freon, trifluorothanol or perfluoro octane) at ambient temperature or extraction with boiling solvent, about 25 % of the polymer remaining on the surface after an initial short washing procedure (3 minutes) could be removed.

When samples were exposed to polar low molecular weight compounds such as water, alcohols, acetone or chloroform and were afterwards washed in a solvent for the polymer such as Freon, trifluoroethanol or FC–72 (a mixture of polyfluorinated hydrocarbons) nearly half of the attached polymer could be removed from the surface as can be seen in figures 2 and 3. Comparable results were obtained upon aging of the samples. Samples which were rinsed with solvent after thermal treatment and kept under ambient conditions for one month lost about one third of their thickness in a subsequent solvent wash.

However it was not possible to remove all of the polymer by displacement even under drastic conditions (boiling in water for 24 – 36 hours). Complete removal of the perfluoropolyether could be achieved only by boiling in mineral acids or –bases for >30 min. However this led also to corrosion of the surfaces which were studied.

In order to investigate how strongly the perfluoropolyethers 1 and 2 are physisorbed and to estimate how much energy is required to remove such material from the surface, the heat of adsorption and the heat of dissolution of the polymers in various fluorinated solvents were determined by microcalorimetry. In order to have a large and well–defined surface the measurements of the heat of adsorption were performed using silica powder with a large surface area (200 m^2/g by BET). The results of the micro-calorimetric results are depicted in table 3. Experimental details and a more detailed discussion of the results will be published elsewhere (11).

The measured heat of adsorption of 1 (ZDOL) on silica (65 kJ/mol) was much smaller than that of water (200 kJ/mol) (12). Therefore it can be inferred that the effectiveness of water in displacing physisorbed PFPE from a carbon surface reflects its higher heat of adsorption to a carbon surface relative to PFPE. Other low molecular weight compounds such as alcohols, which have comparably strong interactions with the surface would also be expected to displace the PFPE. It is interesting to note that the heat of adsorption of the polymer 1 (ZDOL) with hydroxyl endgroups is twice as

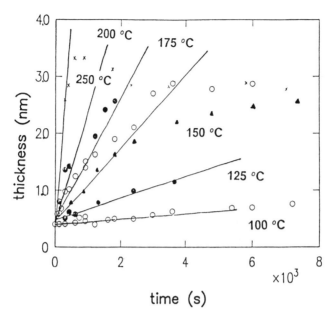

Fig. 1:   Thicknesses of attached polymer films of **1** (ZDOL) on SiO_2 as a function of reaction time at 100, 125, 150, 175, 200 and 250 °C; initial film thickness 10 nm; after thermal treatment (60 min, 150 °C) and extraction with Freon; solid line: initial reaction rate

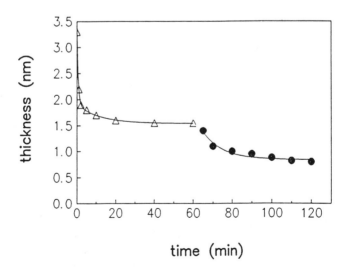

Fig. 2:   Thickness of an attached perfluoroether film after solvent (△) and water exposure (●)

large as that of the polymer 2 with trifluoromethyl endgroups because of the much stronger interaction of the hydroxyl groups with the polar surface groups.

Table 3: Heat of adsorption and heat of dissolution of the PFPE 1 and 2

| | Heat of adsorption[1,2] | Heat of dissolution[1] in: | | |
		Freon	CF_3OH	C_8F_{18}
ZDOL 1	-64 ± 5	10.1	4.5	6.7
Z03 2a	$-33 + 5$	10.0	0.2	5.40
Z15 2b	$-33 + 5$	–	–	–
water	-200^3	–	–	–

[1] in J/g; $\pm$ 0.5 J/g

[2] on silica powder (Aerosil, Degussa, 200 m²/g by BET)

[3] according to ref. (12)

As depicted schematically in fig. 4 the removal of PFPE–lubricant occurs not because the polymer is very soluble in the solvent but rather because displacement occurs. The displaced polymer is not interacting with the surface any more and is readily washed away in a subsequent short exposure to a solvent for the polymer. The time required for displacing the polymer from the surface is therefore determined by the time required for the diffusion of the polar compound through the polymer film. This time is influenced by the film thickness and the amount of polar low molecular weight displacer molecules initially present in the film or on the surface. Thus differing results on comparable surfaces may sometimes be attributed to a different history of the samples e.g. solvent exposure or exposure to water vapor from the ambient.

That the polymer does not gain much energy through solvation with the solvents employed can be concluded from measurements of the heat of solution of 1 and 2 in Freon and other fluorinated solvents, again measured by microcalorimetry. The results of these investigations are shown in table 3. The heats of solution of ZDOL and Z03 in Freon are slightly endothermic (about 10 J/g), which is quite usual for polymers above their glass transition temperature (15). It indicates, that the interactions between the solvent molecules themselves are larger than the interaction between the polymer and the solvent. Thus to dissolve polymer adsorbed onto the

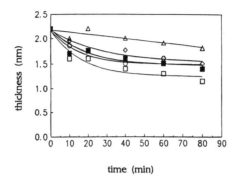

Fig. 3:   Film thickness of 1 (ZDOL) after exposure to various low
molecular weight displacers and subsequent rinsing with
1,1,2–tri-chloro trifluoro ethane; displacer:(△) toluene
(◇) chloroform (○) acetone (■) methanol (□)water

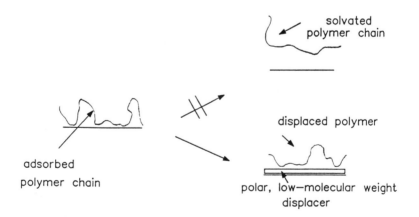

Fig. 4:   Schematic picture of polymer displacement

surface, the energy of adsorption and the energy of solution both have to be overcome.

In order to investigate whether the dilution of the polymer solution is also accompanied by release or consumption of energy the heat of solution was measured in a wide range of final polymer concentrations (0.1 – 33 %). It was found, that the heat of solution was indeed independent of the final concentration in the measured range.

Furthermore it must be considered that not only the nature of the molecules which displace the polymer from the surface is of importance, but also the polarity of the surface to which the polymer is attached plays an important role. This can be seen when the attachment of 1 (ZDOL) to two carbon surfaces with a different surface chemistry is examined.

One polymer film was prepared on a carbon film sputtered on a silicon substrate and a second polymer film was prepared on pyrolytic graphite. Both the sputtered carbon film and the graphite were investigated with XPS–analysis after surface derivatization as described above. The sputtered films contained a large number of hydroxyl–, carboxylic acid and other functional groups at the surface. The pyrolytic graphite sample prepared by freshly cleaving along the basal plane had only a very small number of functional groups on the surface. Both samples were covered with a thick layer of polymer and heated for 90 minutes to 150 ^{0}C. After careful extraction with Freon the film thicknesses of the attached polymer films were determined by XPS–analysis of the fluorine signal. It was found that in both cases about 2.0 nm of 1 was attached during the heat treatment. Control samples prepared under identical conditions, but kept at ambient temperature for one hour gave film thicknesses of only 0.2 – 0.3 nm after solvent extraction according to XPS measurements.

The influence of the surface chemistry can be seen quite clearly if the two samples are exposed to low molecular weight displacers such as water. Both the samples initially had a very high contact angle against water (about 110^0). It was observed for the graphite sample that the water contact angle decreased very rapidly. In fact the spreading of the water droplet was so fast that an exact measurement was difficult to perform and within a few minutes time the contact angle of pure graphite was reached. In contrast the water contact angle of the film of 1 attached to a surface of sputtered carbon changed only very slowly. The wetting angle decreased only a few degrees over a period of several hours.

The results obtained can be attributed to the different surface chemistries of the samples, which can be expected to give different reactivities for chemisorption of the polymers. The sputtered carbon sample contains a large number of reactive groups on the surface compared to the graphite sample and consequently more perfluoropolyether can be chemisorbed to the surface. As chemisorbed polymer is not easily displaced from the surface, the water contact angle stays nearly constant for the sputtered carbon sample. The film attached to the graphite sample is largely physisorbed polymer, which is rapidly displaced by water, and so the water contact angle changes with time.

Subsequent measurements of the film thickness of the polymer attached to a graphite surface by ellipsometry showed that after exposure to displacing molecules the film was no longer homogeneous, but large variations of the film thickness such as bare spots were observed.

Multiple Modes of Atttachment

The results of experiments in which the polymer is attached to different substrates and where the conditions under which polymer is removed from the surface were varied, show quite clearly that there are different types of attachment of the polymer molecules to the surface. Some polymer can be readily washed away. An additional fraction can only be removed by extended solvent extraction (requiring ultrasonics in some cases). Finally, in the case of the hydroxyl–terminated PFPEs there appears to be a fraction which is chemically bonded to the surface. A schematic picture of the different attached polymer molecules is shown in fig. 5. The different mechanisms of attachment cause a greatly varied strength of interaction with the surface and therefore a broad distribution of energies which are required to remove the polymer molecule from the surface. More weakly physisorbed material is displaced readily from the surface; more strongly attached polymer is displaced correspondingly more slowly.
It should be noted that the weakly attached polymer is not necessarily only physisorbed polymer, but may also be polymer which is trapped in the film by entanglement. As discussed in ref. (14) the molecular weight of the polymers 1 and 2 is well below the critical molecular weight for entanglement of polymer chains. However, the broad molecular weight distribution of the polymer and the number weight average increase due to evaporation of low molecular weight polymer must also be taken into account. Furthermore it has to be considered that in such thin films the polymer molecules are all very close to the surface and are thus much more constrained in their motions. Therefore it can be expected that this proximity to the surface will make disentanglement much more difficult for simple steric reasons.
Although we have no direct evidence for entanglement in these films, we cannot *a priori* exclude the possibility.

AFM Measurements

It can be expected that the variation of the interactions of the polymers with the surface will subsequently cause differences in the mobility of the polymer molecules in the films. This mobility can be studied indirectly using the atomic force microscope (AFM). In these experiments samples with thin films of the perfluoropolyether 1 (ZDOL) were attached to the natural oxide of silicon and the weakly attached polymer was removed using different cleaning procedures. When the polymer is only physisorbed to the surface ("free lubricant") it can be observed that at a certain distance between the tip of the AFM–cantilever and the sample, an attractive force suddenly occurs and the tip is dragged closer to the surface (16). This has been attributed to the formation of a meniscus of the 'liquid–like' polymer around the tip (16). This behavior is depicted in the uppermost curve of fig. 6. The attractive meniscus force increases gradually until the sample and tip come into intimate contact and the force becomes repulsive.

This behavior, attributed to the meniscus formation of the physisorbed polymer, has to be compared with samples where the polymer was attached by a thermal treatment (1hr, 150 ^{0}C) and the weakly or non–attached polymer was removed afterwards by dipping it in solvent (Freon) for three minutes. The force–distance diagram still shows some onset of an attractive force at about 5.0 nm due to full or partial formation of a meniscus, but – compared to a sample with physisorbed polymer only – this force is much weaker. Furthermore the transition is much less defined than in samples with physisorbed polymer only, it is smeared out over a distance of some 0.5 nm.

In a second series of samples the weakly physisorbed (or entangled) polymer was removed by cleaning the sample after thermal treatment with Freon using ultrasound or by displacing the polymer with water. The solvent extracted sample was immersed in water for 18 hours followed by additional freon cleaning. The thicknesses of the polymer films were 1.2 and 1.7 nm respectively according to ellipsometry. These samples showed even more pronounced differences in the force distance behavior when compared to samples with physisorbed polymer only. When all of the weakly adsorbed polymer is displaced and subsequently dissolved, reducing the tip–sample distance under 4.0 nm causes only a very slow increase of the attractive force. No abrupt increase in the attractive force due to a meniscus formation was observed. This different behavior can be explained in the fact that most of the mobile lubricant was removed by displacement and dissolution through the exposure to displacers or the extended solvent cleaning of the samples. Only firmly attached polymer remains and this is immobile, at least on the time scale of these measurements. Therefore these results also support a model in which several types of attachment of the perfluoropolyether molecules are coexisting in the films.

Conclusions

This study has shown that in ultrathin films of end–group functionalized perfluoropolyethers attached to materials such as carbon, aluminum or SiO$_2$ the strength of attachment of the individual polymer molecules to the surface varies greatly and different modes of attachment coexist. The films contain both physisorbed (or entangled) and chemisorbed material. Their relative amounts depend on the surface chemistry of the substrate and the history of the sample. Temperature, length of the heat treatment and exposure to solvents and polar displacers are all important factors in determining the nature of attachment.

Microcalorimetry measurements demonstrate that physisorbed perfluorinated polymers 1 and 2 are easily displaced by water or other polar low molecular weight compounds independent of the nature of the endgroup. The displacement experiments however show that chemisorbed polymer cannot be so easily removed.

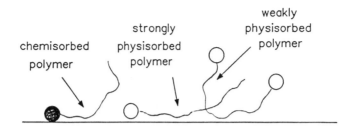

Fig. 5:   Different attachment mechanisms

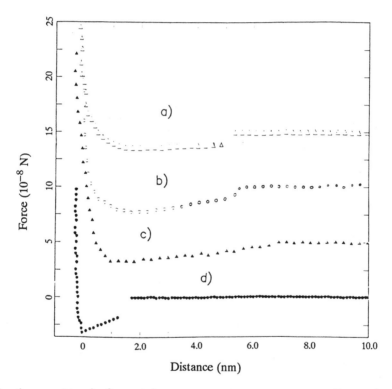

Fig. 6:   Atomic force microscopy measurements on perfluoroether films with different strength of attachment: a) physisorbed polymer b) after thermal treatment and solvent exposure c) after displacement with water and subsequent Freon extraction d) clean silicon; measurements performed under ambient conditions

Measurements with atomic force microscopy also support a model in which different modes of attachment coexist. If physisorbed polymer is dissolved or displaced from the surface only chemisorbed polymer remains, thereby reducing the mobility of the polymer in the film dramatically.

The results obtained emphasize that a detailed characterization of a deposited polymer film is necessary prior to evaluation of the tribological behavior of such samples. Otherwise slight changes in the treatment of the samples will give different amounts of physisorbed and chemisorbed material, resulting in different tribological properties.

Acknowledgements:

The authors wish to thank J. Burns, IBM, Storage System Products Division, San Jose, for the NMR–measurements of the perfluoropolyethers.

References:

1) A. Homola, M. Mate and G.B. Street, MRS Bull. 15, 45 (1990)
2) B. Bhushan, Wear, 136, 169 (1990)
3) F.P. Bowden and D. Tabor, The Friction and Lubrication of Solids, Clarendon, Oxford, (1954)
4) E. Ando, Y. Goto, K. Morimoto, K. Ariya and Y. Okahato, Thin Solid Films 180, 287 (1989).
5) V. Novotny, J.D. Swalen and J.P. Rabe, Langmuir 5, 485 (1989)
6) J. Seto, T. Nagai, C. Ishimoto and H. Watanabe, Thin Solid Films, 134, 101 (1985)
7) A.M. Scarati and G. Caporiccio, IEEE Trans. Magn. Mag. 23, 106 (1987)
8) M.Suzuki, Y. Saotome and M. Yanagisawa, Thin Solid Films 160, 453 (1988)
9) T. Miyamoto, I. Sato and Y. Ando in: Tribology and Mechanics of Magnetic storage systems Vol.5, B. Bushan and N.S. Eiss (eds.) (1988) p. 55
10) D. Sianesi, V. Zamboni, R. Fontanelli, M. Buraghi, Wear 18, 85 (1971)
11) J. Rühe, S. Kuan, J. Burns, T. Clarke and G.B. Street, to be published
12) R. K. Ihler, The Chemistry of Silica, J. Wiley, New York (1979) p. 692
13) J. Rühe, S. Kuan, V. Novotny, T. Clarke and G.B. Street, to be published
14) V. Novotny, J. Chem. Phys. 92, 3189 (1990)
15) R. Orwool in: Polymer Handbook, J. Brandup and E.H. Immergut (eds.), Wiley, New York (1989) p. VII–517
16) C.M. Mate, M.R. Lorenz and V.J. Novotny, J. Chem. Phys. 90, 7550 (1989)

RECEIVED December 3, 1991

Chapter 10

The Decomposition Mechanism of Perfluoropolyether Lubricants during Wear

Gerard Vurens[1], Raymond Zehringer[1], and David Saperstein[2]

[1]Almaden Research Center, IBM Research Division, 650 Harry Road, San Jose, CA 95120–6099
[2]IBM Storage Systems Products Division, 5600 Cottle Road, San Jose, CA 95193

The decomposition of perfluoropolyether (PFPE) lubricants is studied by Fourier Transform Infra Red Spectroscopy (FTIR), Thermal Programmed Desorption (TPD), Electron Stimulated Desorption (ESD) and mass spectrometry. Thermal decomposition of perfluoropropyleneoxide (Krytox, Dupont) results in a different set of products than electron decomposition. The electron decomposition of perfluoropolyethers can take place with electrons with energies below the ionization potential of the perfluorinated molecule. This is probably due to the formation of a negative ion resonant state, followed by the dissociation of the molecule into a negative ion and a radical. Mechanical decomposition gives rise to a third product distribution, that bears some similarity to the electron decomposition products.

Mechanical energy can be used to initiate chemical reactions. Carey-Lea [1, 2, 3] showed in the last century how mercury chloride could be decomposed into its elements by rubbing in a mortar. Since mercury chloride sublimes without decomposing when it is heated, frictional heating could be excluded as the cause of the decomposition. Besides mercury chloride there are many substances that decompose under the influence of mechanical energy. An excellent review is written by Heinicke [4]. In some instances the decomposition has been explained by rapid local heating during impact between two materials. In other occasions frictional heating could be excluded. During the sixties Thiessen et al. [5] developed the *magma plasma model* in order to clarify contradictory opinions. The *magma plasma model* assumes that mechanical excitation initiates an excited state (a plasma) of the material, wherein its lattice components and free electrons become separated. Thiessen et al. distinguishes between several different groups of chemical reactions that can subsequently occur. The first group (I) are the reactions taking place in the plasma. The second group (II) are the reactions taking place during dissipation of the plasma energy (for instance reactions occurring due to

0097–6156/92/0485–0169$06.00/0

high local temperatures), and the third group (III) of reactions are somehow catalyzed by the energy stored in the solid; i.e. defects etc. During the last twenty years the development of modern surface science techniques has enabled people to study chemical reactions occurring on the surfaces of solids[6]. Oftentimes the reaction studied was catalyzed by these solid materials. Therefore there is a lot of information about the reactions from group III. It is known that the generation of defects in a material can accelerate reactions [7, 8, 9, 10].

In this paper we studied the decomposition of perfluoropolyether (PFPE) lubricants by means of mass spectrometry and Fourier Transform Infra Red spectroscopy. Thermal decomposition of perfluoropolyethers has been studied previously [11, 12]. High energy electron decomposition has also been studied by several groups [13, 14, 15, 16, 17]. Other ways of degrading perfluoropolyethers such as x-ray degradation [18] and γ-ray degradation [19] have been reported. The products being generated during mechanical decomposition are compared with products from thermal decomposition and from electron decomposition of these perfluoropolyethers. Perfluoropolyethers like all perfluorinated compounds are very susceptible to decomposition by electrons that might be generated in the tribo plasma. On the other hand local high temperatures could decompose the perfluoropolyethers too. Evidence will be shown that mechanical decomposition of perfluoropolyethers is an electron mediated process.

Experimental

The mass spectrometry, including the Electron Stimulated Desorption (ESD) and Thermal Programmed Desorption (TPD), was performed in a different vacuum system than the Fourier Transform Infrared Spectroscopy (FTIR). In order to distinguish between products being generated by thermal decomposition and by electron decomposition with the mass spectrometer, thin films (about 20Å) of perfluoropolyethers on silicon wafers were bombarded with electrons from an electron gun (Apex Electronics model 3K-5U) in an electron stimulated desorption experiment. A typical filament bias voltage was 10 V and a typical sample current was 10 pA. The experiments were performed in a vacuum system with an operating pressure of approximately 1×10^{-8} Torr. The fragments being generated were detected with a quadrupole mass spectrometer (Model EXM-280, Extrel Corp.). This mass spectrometer has an independently operated ionizer, that can be turned off in order to detect externally generated ions. This mass spectrometer allowed us to scan between 0 and 350 mass units every 5 seconds resulting in a complete mass spectrum of the products with a 5 second time resolution. In order to obtain good signal to noise the spectra were averaged over several (typically two) minutes. Before electron bombardment is started the filament is heated at a positive potential and background spectra are taken and averaged. To start electron bombardment the filament potential is reversed to -10 V. After several minutes of spectral averaging the background spectrum is subtracted. The TPD experiments were performed similarly. A thin film of perfluoropolyether is deposited on a tantalum foil that is resistively heated. Before the heating is started a background spectrum is taken and the product generation is followed with a 1 second time resolution. Then the temperature is ramped up to 700 °C at a rate of approximately 1 °C/s. During this ramp the product distribution is measured with a 1 second time resolution. The mechanical decomposition products were generated by rubbing a 1 millimeter diameter sapphire ball onto an amorphous carbon overcoated spinning disk with a load of roughly 5 g. Again a background spectrum was taken before the load was applied. After the load was applied the generated products that expanded into the vacuum system were ionized and detected by the mass spectrometer.

Table 1: The different perfluoropolyethers used in this study and their structure

Tradename/ Supplier	Structure
Fomblin Z15/ Montedison	CF_3O-$(CF_2CF_2O)_n$-$(CF_2O)_mCF_3$
Demnum S200/ Daikin	$F(CF_2CF_2CF_2O)_n$-CF_2CF_3
Krytox 143 AD/ Dupont	$F(CF(CF_3)CF_2O)_n$-CF_2CF_3

The FTIR data was taken with an FTIR spectrometer (IBM instruments model 44). For the FTIR experiments approximately 200 mg of perfluoropolyether lubricant was heated in an evacuated cell. The cell was evacuated before the experiment and was subsequently isolated from the pump. The initial pressure in the cell was approximately 1×10^{-7} Torr. The cell was made of stainless steel and was equipped with 2 high current electrical feedthroughs for the filament (electron source) and two 25 mm diameter sodium chloride windows for the FTIR transmission experiments. The temperature was measured using a thermocouple that was connected to the outside of the vacuum cell. The FTIR spectrum was then taken of the accumulated gas phase products.

In order to detect the electron decomposition products in the FTIR experiments, the perfluoropolyether samples were decomposed by electron bombardment from a hot filament inside the evacuated cell. The hot filament caused partial thermal decomposition of the generated gas phase products.

Results and Discussion

The experiments were performed using several different perfluoropolyethers listed in table 1. A FTIR study was made to compare the thermal decomposition of Fomblin Z15 (Montedison), Demnum S200 (Daikin Industries) and Krytox 143 AD (Dupont) with the electron decomposition of Fomblin Z15 and Demnum S200. The mass spectrometry studies compared the thermal and electron decomposition of Krytox 143 AD with the decomposition caused by wear.

FTIR. Figure 1 shows representative FTIR spectra of the thermal decomposition of Fomblin Z15 (a), Demnum S200 (b) and Krytox 143 AD (c). All perfluoropolyethers decomposed at elevated temperatures and formed a variety of gas phase products, such as CO, CO_2, COF_2 and CF_4. F_2 if formed is not infrared active and cannot be measured. Fomblin Z15 gave rise products at significantly lower temperature than Demnum S200 or Krytox 143 AD. Unexpected decomposition products of Fomblin Z15 that were detected included HF and SiF_4. The formation of HF is surprising since the Fomblin Z15 does not contain hydrogen atoms. Moreover some HF is formed during the decomposition of all perfluoropolyethers in vacuum. Apparently enough water is adsorbed on the walls of the vacuum system to form a significant amount of HF, by reaction with the fluorinated products. Due to the timescale of the FTIR experiment (approximately 30 min) the more reactive gas phase products of the perfluoropolyethers can react with the water on the walls to form HF. The hydrogen fluoride subsequently reacted with the silicon oxide of a ceramic feedthrough in the vacuum cell to form SiF_4. This was concluded from the decrease of HF in time coinciding with an increase in

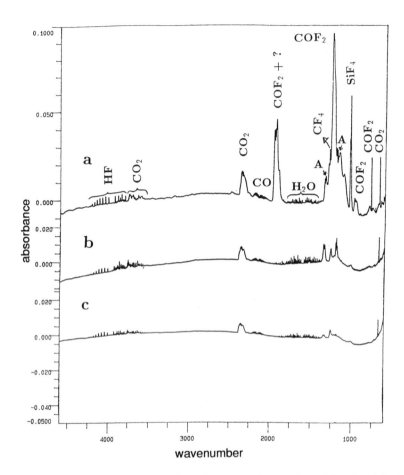

Figure 1. The thermal gas phase degradation products of Fomblin Z15 (a), Demnum S200 (b) and Krytox 143 AD (c) as measured by FTIR. Note the change in scale for figure (a) on the absorbance axis.

the SiF_4 signal which obviously is not a direct product of the perfluoropolyether decomposition.

The water that is detected by the FTIR could come from degassing from the walls of the vacuum cell during heating or from fluctuations in the water vapor concentration in the nitrogen surrounding the vacuum cell in the FTIR spectrometer. Similarly CO_2 could be due to fluctuations of the CO_2 concentration in the nitrogen surrounding the vacuum cell, but it is unlikely since the CO_2 signals scale with the approximate total amount of other products formed. It is concluded that CO_2 is a stable end product of the perfluoropolyether decomposition. The origin of the CO could be from the decomposition of the perfluoropolyethers or from outgassing of the stainless steel walls of the vacuum cell during heating. However we deduce that CO is a real decomposition product since the cell was heated to a much higher temperature ($\approx 600\,°C$) in the case of Demnum S200 and Krytox 143 AD than in the case of Fomblin Z15 ($\approx 450\,°C$), while the formation of CO is largest in the case of Fomblin Z15 (compare figures 1a and 1b). Finally there are at least 3 other products resulting from the thermal decomposition of Fomblin Z15. These are COF_2, CF_4 and an unknown product, which we called **A**. The same product **A** results from the thermal decomposition of Demnum S200 and from Krytox 143 AD, but less is formed (figures 1b and 1c respectively). Also in figure 1a additional peaks in the region of $1950\,cm^{-1}$ are observed suggesting yet another unknown product. Due to the absence of a C-H stretch between 3000 and $3200\,cm^{-1}$ in the FTIR spectra it is assumed that these unknown compounds entirely exist of carbon, oxygen and fluorine.

Figure 2 shows the electron decomposition results of Fomblin Z15 as measured with FTIR. Figure 2a shows that there is a small amount of CO degassing from the filament, when it is positively biased. Switching the bias from positive to negative gives rise to a number of electron bombardment decomposition products (figure 2b). These include HF, CO_2, CO, COF_2, CF_4, SiF_4 and possibly CHF_3. The identification of fluoroform (CHF_3) is in doubt, because the expected CH stretch could not be detected, even though a sizable $1152\ cm^{-1}$ peak was observed. After switching the filament potential back to positive, another FTIR spectrum was recorded (figure 2c). From the difference in figures 2b and 2c it is concluded that COF_2 is not a stable product in the presence of a hot tungsten filament.

The same set of experiments, but now with Demnum S200 is summarized in figure 3. Figure 3a shows the initial outgassing of the tungsten filament. Figure 3b shows the gas phase electron decomposition products of Demnum S200. It consists of the same products as the Fomblin Z15 electron decomposition products. Continued heating of the filament at positive potential decomposes the COF_2 that was formed. It should be noted that COF_2 is not one of the Demnum S200 thermal decomposition products. This is strong evidence that the electron decomposition and thermal decomposition processes of perfluoropolyethers are fundamentally different and distinguishable.

Although the FTIR results are encouraging, they clearly show the need for a measurement technique with a higher sensitivity, so that less material could be detected in a shorter time. This could prevent the formation of secondary products such as HF and SiF_4. Quadrupole mass spectrometry was used to obtain high speed and high sensitivity product detection.

Mass Spectrometry. D'Anna et al [20] showed that the ionization potential for perfluoropolyether materials is approximately 14 eV, a number which is typical for many perfluorinated compounds. However it is possible to decompose

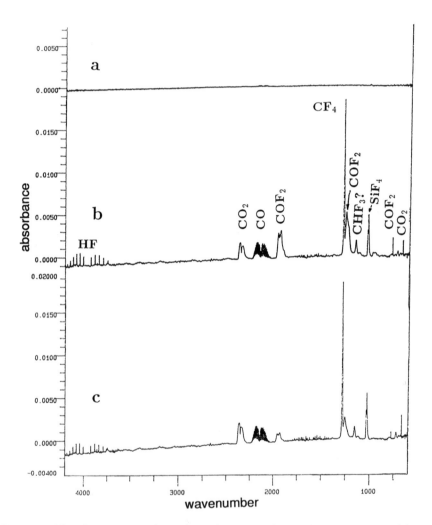

Figure 2. The electron gas phase degradation products of Fomblin Z15. (a) is the filament at a positive bias, (b) after electron bombardment and (c) after subsequently heating the filament at a positive bias.

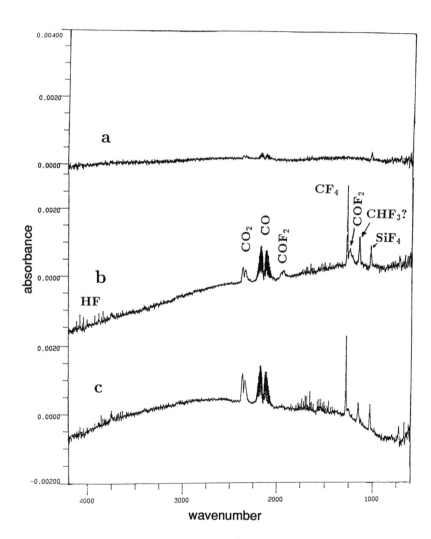

Figure 3. The electron gas phase degradation products of Demnum S200. (a) is the filament at a positive bias, (b) after electron bombardment and (c) after subsequently heating the filament at a positive bias.

perfluoropolyethers with electrons of energies less than 14 eV. In this case neutral species and negative ions will be emitted from the sample. Figure 4 shows the negative ions emitted from Demnum S200 during irradiation with low energy electrons, i.e. less than 14 eV. The most intense signal (80% of the total integrated ion current) is from F^- at $m/e = 19$. This is expected since F^- is a very stable ion. Other ions that show a strong signal are $C_3F_5O_2^-$ and $C_3F_7O^-$. at $m/e = 163$ and $m/e = 185$ respectively. These two ions are characteristic for the Demnum monomer unit.

The fragmentation pattern of the neutral species being emitted from the surface during electron bombardment is shown in figure 5a for the case of Krytox 143 AD. By far the largest peak is at $m/e = 69$ (CF_3^+), followed by peaks at 28 (CO), 44 (CO_2), 20 (HF) and 119 ($CF_3CF_2^+$). No peaks were detected in the range from 200 to 350 amu. This suggests that only low molecular weight products are being formed. The detected products agree well with the products found in the FTIR experiment. SiF_4, (major peak at $m/e = 85$ (SiF_3^+)) was not found and CHF_3 could not be confirmed. There is a small peak at $m/e = 66$ (COF_2) but this could also be due to fragmentation of a higher molecular weight product. Thus COF_2 formation can not be uniquely identified as a direct electron decomposition product. The presence of $m/e = 69$ could be traced directly to Krytox, but $m/e = 69$ is also the major fragmentation peak of CF_4 and many other perfluorocarbons. The fragmentation pattern of the products from the thermal decomposition of Krytox 143 AD are shown in figure 5b. The decomposition is maximum at 450 °C. The strongest peak in the spectrum is 169 ($C_3F_7^+$), followed by 69 (CF_3^+), 119 ($C_2F_5^+$), 147 ($CF_2CF_2COF^+$), 97 (CF_2COF^+) and 150 ($C_3F_6^+$) respectively. In addition high m/e peaks were found at $m/e = 219$ ($C_4F_9^+$) with intensity 0.6, $m/e = 285$ ($C_5OF_{11}^+$) with intensity 4.4 and $m/e = 335$ ($C_6OF_{13}^+$) with intensity 0.3. The intensities are normalized with respect to $m/e = 169$ which has intensity 100. The C5 and C6 fragments are probably dimeric fragments. A possible dimeric product formed by scission of the Krytox main chain is perfluoro-iso-propyl-n-propylether. Upon ionization the molecule has 3 possible ways of dissociating into a CF_3 radical and a positive ion of $m/e = 285$.

Electron stimulated desorption appears to be a powerful characterization tool for the determination of lubricant structure of thin films. Secondary Ion Mass Spectrometry (SIMS) [21, 22, 23] and X-ray Photoelectron Spectroscopy (XPS) [18] have also been used for this type of characterization. Emission of products were detected with electrons as low as approximately 2 eV kinetic energy [24]. A review of the decomposition of perfluorinated compounds by low energy electrons has been written by Christophorou [25]. Low energy electrons readily attach to perfluorinated compounds, forming a negative ion resonance state. The negative ion resonance state then causes the dissociation of the molecule into radical and negative ion fragments both of which can be detected with a mass spectrometer. This process is called dissociative electron attachment. Because of the ease with which perfluoropolyethers decompose with low energy electrons, it suggests that their decomposition takes place through the formation of a negative ion resonance state.

The mass spectrometry data confirms the FTIR data in that a different set of products is generated during electron bombardment than during thermal decomposition. Thermal decomposition leads to higher molecular weight products than electron decomposition. Other differences include a greater fraction of oxygenated products during thermal degradation as shown by the ratio of the peaks 97/100 and 147/150. These ratios and the molecular weight of the products (for instance

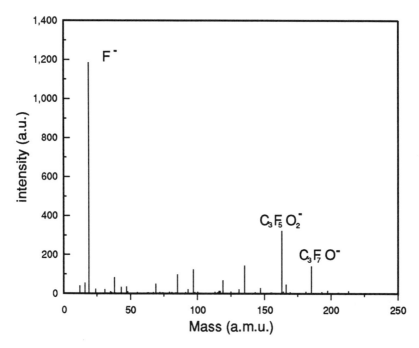

Figure 4. The negative ions emitted from Demnum S200 upon irradiation of low energy electrons.

by taking the 69/169 ratio) can be used as a signature for electron decomposition and thermal decomposition of Krytox.

Figure 5c shows the mass spectrum of the gas phase products generated during wear of a thin film of Krytox 143 AD on amorphous carbon. The largest peak in the spectrum is at m/e = 85 (CF_3O^+). The rest of the spectrum is strikingly similar to the low energy electron decomposition products. This is clearly shown by the intensities of the m/e = 69, 119, 169 (CF_3^+, $CF_2CF_3^+$, $CF_2CF_2CF_3^+$ respectively) series and by the 97/100 peak ratios. The peak at m/e = 85 could also be attributed to SiF_3^+, the major fragmentation peak of SiF_4, but no silicon containing materials were present at the interface during wear. A sapphire (Al_2O_3) ball was used to induce wear of the carbon. It is therefore concluded that the fragment at 85 amu is indeed CF_3O^+, a mechanical decomposition product of Krytox 143 AD. This product is not observed during either electron or thermal decomposition of Krytox 143 AD.

Conclusions

The results clearly show that electron decomposition of perfluoropolyethers gives rise to a different set of products than thermal decomposition. This shows that electron decomposition and thermal decomposition of these perfluoropolyethers probably occurs through different pathways. It has been shown that thermal decomposition of perfluoropolyethers can proceed through a catalytic mechanism involving Lewis acid centers [11, 12, 26]. Our data suggests that the electron decomposition occurs through the formation of a negative ion resonance state.

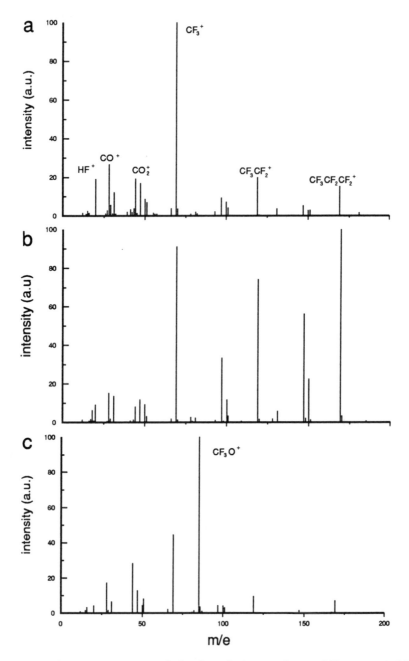

Figure 5. The mass spectrum of the degradation products of Krytox 143 AD after electron bombardment(a), after heating to 450 °C (b) and during wear (c).

Then a negative ion is formed through dissociative electron attachment and a reactive radical species is generated. Subsequently gas phase products are generated from the reactive radical species as well as from the negative ions. FTIR shows that both the products of electron and thermal decomposition readily react with water to form hydrogen fluoride. Hydrogen fluoride is a very corrosive gas, that can react with many ceramics and glasses producing a variety of secondary products. It is clear that mass spectrometry is sensitive enough to detect small amounts of wear fragments from these perfluoropolyethers. The wear fragments suggest that the decomposition of the perfluoropolyethers is an electron mediated process, since most of the product fragment distribution is very similar to the electron decomposition fragment distribution. The electrons necessary for this decomposition could be triboelectrons generated in the tribo plasma as described by Thiessen [5]. A prevalent mechanical decomposition product fragment at $m/e = 85$ (CF_3O^+) could not be explained by either a thermal or an electron mediated mechanism. The origin of this product is currently under investigation.

Acknowledgements

The authors gratefully acknowledge the help of Judy Lin and Leela Viswanathan in the preparation of the samples and Chris Gudeman and John Foster for the helpful discussions.

References

[1] M. Carey-Lea. *Phil. Mag.*, 37:31, 1894.

[2] M. Carey-Lea. *Phil. Mag.*, 34:46, 1892.

[3] M. Carey-Lea. *Phil. Mag.*, 37:470, 1894.

[4] G. Heinicke. *Tribochemistry.* Carl Hanser Verlag, 1984.

[5] P. A. Thiessen, K. Meyer, and G. Heinicke. *Grundlagen der Tribochemie*, volume 1 of *Abh. Dtsch. Akad. Wiss., Kl. Chem. Geol. u. Biol.* 1966.

[6] G. A. Somorjai. *Chemistry in two Dimensions: Surfaces.* Academic Press, 1981.

[7] M. Salmeron, F. J. Gale, and G. A. Somorjai. *J. Chem. Phys.*, 70:2807, 1979.

[8] T. E. Madey, J. T. Yates, D. R. Sandstrom, and R. J. M. Voorhoeve. volume 6B. Plenum Press, New York, 1976.

[9] J. H. Lunsford. *J. Phys. Chem.*, 68:2312, 1964.

[10] M. Boudart, A. Delbouille, E. G. Derouane, V. Indovina, and H. B. Walters. *J. Am. Chem. Soc.*, 94:6622, 1972.

[11] L. S. Helmick and W. R. Jones. *NASA Tech. Mem.*, 102493, 1990.

[12] M. J. Zehe and O. D. Faut. *NASA Tech. Mem.*, 101962, 1989.

[13] M.A. Baker, L. Holland, and L. Laurenson. *Vacuum*, 21:479, 1971.

[14] L. Holland, L. Laurenson, P.N. Baker, and H.J. Davis. *Nature*, 238:36, 1972.

[15] L. Holland, L. Laurenson, R.E. Hurley, and K. Williams. *Nucl. Inst. Meth.*, 111:555, 1973.

[16] J. Pacansky, R.J. Waltman, and M. Maier. *J. Phys. Chem.*, 91:1225, 1987.

[17] J. Pacansky and R.J. Waltman. *J. Phys. Chem.*, 95:1512, 1991.

[18] F. Pan, Y. Lin, and S. Horng. *Appl. Surf. Sci.*, 47:9, 1991.

[19] P. Barnaba, D. Cordischi, A. Delle Site, and A. Mele. *J. Chem. Phys.*, 44:3672, 1966.

[20] E. D'Anna, G. Leggieri, A. Luches, and A. Perrone. *J. Vac. Sci. Technol.*, A5:3436, 1987.

[21] I.V. Bletsos, D.M. Hercules, D.E. Fowler, D. vanLeyden, and A. Benninghoven. *Anal. Chem.*, 62:1275, 1990.

[22] D.E. Fowler, R.D. Johnson, D. vanLeyden, and A. Benninghoven. *Anal. Chem.*, 62:2088, 1990.

[23] D.E. Fowler, R.D. Johnson D. vanLeyden, and A. Benninghoven. *Surf. Int. Anal.*, in press, 1991.

[24] G.H. Vurens, C.S. Gudeman, L.J. Lin, and J.S. Foster. *In Preparation.*

[25] L. G. Christophorou. *Electron-Molecule Interaction and Their Applications.* Academic Press, 1984.

[26] P. Kasai. *In Preparation.*

RECEIVED November 5, 1991

Chapter 11

The Influence of Fluorination on Boundary-Layer Surface Chemistry

Bryan Parker, Ruiming Zhang, Qing Dai, and Andrew J. Gellman*

Department of Chemistry, University of Illinois, Urbana, IL 61801

We have studied the chemistry of model fluorinated boundary layer films adsorbed on the Ag(110) surface. The compounds studied in this work are the ethanols (CH_3CH_2OH and CF_3CH_2OH) and the acetic acids (CH_3CO_2H and CF_3CO_2H) which are subsets of a series of straight chain compounds that have been examined on this surface. This investigation has made use of desorption measurements and surface vibrational spectroscopy to study these molecules adsorbed on both clean and oxidized surfaces. Whereas both ethanols adsorb reversibly on the clean Ag(110) surface they deprotonate to form ethoxides on the pre-oxidized Ag(110) surface. The ethoxides decompose at high temperatures by breaking the C-H bond at C_1 to form acetaldehyde. The fluorinated ethoxide is much more stable than hydrocarbon ethoxide in that it decomposes at 420K as opposed to 270K. We suggest that the initial C-H bond breaking process is one of β-hydride elimination and that the transition state involves formation of a cation at the C_1 position. This cation is destabilized by the increased electronegativity associated with fluorination of the methyl group and hence the fluoro-ethoxide is more stable on the surface. In the case of the acids we find that, whereas acetic acid will adsorb reversibly on the clean Ag(110) surface, 2,2,2-trifluoro-acetic acid deprotonates to form an acetate. The increased electronegativity of the trifluoromethyl group facilitates deprotonation of trifluoro-acetic acid by stabilizing the anionic product.

The boundary layer additives to lubricant mixtures are often species such as carboxylic acids, phosphates, alcohols and other amphiphiles which can adsorb on surfaces from solution to form surfactant-like monolayers. Under boundary layer sliding conditions in which all the lubricant fluid has been excluded from the contact region between two surfaces these monolayer films serve to prevent direct metal-metal contact and hence to reduce friction and wear (*1*). We have

---

*Corresponding author

0097–6156/92/0485–0181$06.00/0

studied a number of these types of species adsorbed on metal surfaces attempting to understand both the surface chemistry and the structure of these adsorbates. Although we are not able to study these materials under contact conditions, by preparing monolayer films of these species on clean surfaces under ultra-high vacuum conditions we are able to apply a number of spectroscopic methods to their characterization.

The materials chosen for study in this work are models for fluorinated compounds which might be used as boundary layer additives to fluorinated fluids used as lubricants at high temperatures. These fluids are perfluoropolyalkylethers (PFAE's) which are stable liquids over very wide temperature ranges but are limited by the lack of potential boundary layer additives (2,3). With the exception of some study of perfluoro-diethylether adsorbed on metal surfaces there has been very little work on the surface chemistry of fluorinated compounds (4).

The aim of the work described here has been to understand the surface chemistry of some small model boundary layer compounds and in particular to compare hydrocarbon amphiphiles with their fluorinated analogues to determine the influence of fluorination on surface chemistry. The model boundary layer compounds are acetic acids and ethanols. Although these are much smaller than any of the long chain amphiphiles which are commonly used as boundary layer additives it has been possible to study their surface chemistry in some detail. Furthermore, they fit our needs in that the methyl group can be fluorinated and hence we are able to examine the influence of fluorination on the surface chemistry of the polar groups of these compounds. The choice of the Ag(110) surface has been made on the basis of the fact that the surface chemistry of the short chain hydrocarbon acids and alcohols has been studied in great detail as a result of their importance in catalytic chemistry. The Ag(110) surface is also relatively passive in that it allows us to isolate both the molecularly adsorbed species and intermediates in the surface reactions. The work described here is part of a larger body of work in which we have looked at the longer chain amphiphiles adsorbed on both Ag and Cu surfaces.

Adsorption of ethanol and acetic acid on both clean and oxidized Ag(110) surfaces has been studied using desorption methods and X-ray Photoemission spectroscopy (5). The primary observation in both cases is that on the clean surface these molecules are adsorbed reversibly. Heating results in ethanol desorption with a heat of desorption of 10.7 kcal/mole (9) and acetic acid desorption with a heat of desorption of 11.5 kcal/mole (6,10). Adsorption of ethanol is thought to be via the lone pair on the oxygen atom. In contrast, on a Ag(110) surface that has been pre-oxidized these molecules are deprotonated to yield the ethoxide or an acetate species. These are much more stable on the surface than the molecular species in that they do not desorb during heating but rather decompose. During heating the ethoxide decomposes at 270K by breaking a C-H bond at the C_1 atom to produce the acetaldehyde which then desorbs rapidly. Other decompostion products include H_2O, C_2H_4, and CH_3CH_2OH. The acetate decomposes at 610K to generate CO_2 and a number of other products including CH_4, CH_3CO_2H, C_2H_2O and surface carbon. Similar behavior has been observed in the cases of methanol (7) and formic acid (8) on these surfaces and more recently in the cases of the longer chain alcohols and acids (9,10). In the case of the longer chain alcohols adsorbed on the clean Ag(110) surface quantitative XPS measurements have been made to determine that these molecules are oriented with their alkyl chains roughly parallel to the metal surface. The heat of adsorption of methanol is 9.7 kcal/mole and for the longer chain alcohols is incremented by $\sim 1.1 \pm 0.1$ kcal/mole/CH_2 group. In the case of the straight chain acids, although we do not have any direct evidence of their orientation on the surface we find that their heats of adsorption are incremented by $0.9 \pm .02$ kcal/mole/CH_2 group (10).

Experimental

The experiments described in this paper were all performed in two ultra-high vacuum chambers. Each chamber was equipped for surface characterization using both Low Energy Electron Diffraction (LEED) and Auger Electron Spectroscopy (AES). In addition the surface preparative methods available include Ar^+ ion sputtering and leak valves for gas dosing. Both chambers were equipped with quadrupole mass spectrometers for desorption measurements and one with a spectrometer for High Resolution Electron Energy Loss Spectroscopy (HREELS).

The Ag(110) crystals were both cleaned by sputtering and exposure to oxygen at high temperatures (500K - 700K). The criterion for surface cleanliness of carbon was that following oxygen adsorption there should be no desorption of CO_2. Production of pre-oxidized surfaces was accomplished by exposing the clean surface to 10L oxygen at 300K. This was sufficient to saturate the surface with atomic oxygen in a p(2x1) lattice. The alcohols and acids used in the work were purified of high vapor pressure contaminants by cycles of freezing and thawing. The alcohols were then introduced into the chamber via a leak valve fitted with a capillary array doser. The acids had to be introduced into the chamber through an effusive pinhole source.

The desorption measurements were made by adsorbing the molecule of interest onto the appropriately prepared surface at T < 130K. The sample was then heated resistively in front of the aperture to a quadrupole mass spectrometer while monitoring several different charge to mass ratios. The temperature was measured using a chromel-alumel thermocouple spotwelded to the sample and was controlled by a computer using a proportional-differential feedback routine to maintain a constant heating rate. The HREELS spectra were obtained at low temperature (T < 120K). The beam energy was 5.5 eV and the spectra were collected in single scans with dwell times of 1 sec/pt.

Results

Ethanol. The adsorption of ethanol on both the clean and pre-oxidized Ag(110) surface has been studied using both desorption measurements and HREELS. The desorption measurements are in agreement with previous studies of both methanol and ethanol adsorbed on these surfaces (5,7) and the vibrational spectra obtained using HREELS confirm the proposed chemistry. On the clean surface ethanol will adsorb reversibly. At low coverages ethanol is bound directly to the surface and desorbs with first order kinetics while at higher coverages, once the monolayer is saturated, ethanol forms a condensed multilayer phase which desorbs with zero-order kinetics. The monolayer desorption spectrum indicates a heat of adsorption of 10.7 kcal/mole and is shown in Figure 1a.

On the pre-oxidized surface ethanol is deprotonated at low temperatures by surface oxygen atoms resulting in the formation of adsorbed water and ethoxide. This same chemistry has been observed before in the case of methanol using desorption measurements (7). Heating the surface results in the desorption of water at 180K leaving the surface covered by the ethoxide layer. This ethoxide monolayer is much more stable than ethanol in the sense that it cannot desorb from the surface. Instead, it decomposes at 270K resulting in the desorption of acetaldehyde as shown in Figure 1b.

The HREELS spectrum of both the reversibly adsorbed ethanol and the ethoxide are both shown in Figure 2. Ethanol is easily identified in Figure 2a by comparison with its liquid phase spectrum whose modes are listed in Table I (11). The spectrum in Figure 2b is that of ethoxide formed by adsorption of ethanol onto a pre-oxidized surface and then heating to 250K to remove water. The spectrum in Figure 2b which contains the same modes as the ethanol vibrational spectrum with the conspicuous absence of the O-H stretch (3300 cm^{-1}) and the O-H out-of-plane bend (680 cm^{-1}). There is some change in the relative intensities of the remaining features which is primarily attributed to changes in orientation of the adsorbate.

2,2,2-trifluoro-ethanol. The chemistry of 2,2,2-trifluoro-ethanol is very similar to that of ethanol adsorbed on the Ag(110) surface. The molecule is adsorbed reversibly with a heat of adsorption of 10.9 kcal/mole. As the coverage is increased from zero the molecule first forms a monolayer which then saturates and further adsorption is in the form of a condensed multilayer. The desorption of trifluoro-ethanol from the clean Ag(110) surface is shown in Figure 3a. As in the case of ethanol the trifluoro-ethanol is deprotonated on the pre-oxidized surface to yield trifluoro-ethoxide and water. The water desorbs during heating at 200K leaving the trifluoro-ethoxide on the surface. The thermal decomposition of the trifluoro-ethoxide leads to the formation and desorption of trifluoro-acetaldehyde as illustrated in Figure 3b. The primary difference between the surface chemistry of ethoxide and its fluorinated counterpart is the kinetics of the decomposition reaction. It is apparent from comparison of Figures 1b and 3b that the fluorinated ethoxide is much more stable than ethoxide, decomposing on the surface at 420K as opposed to 270K.

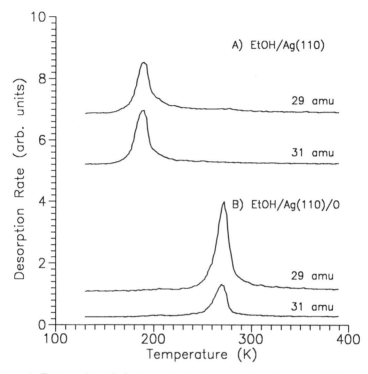

Figure 1. Desorption of a) ethanol adsorbed on the clean Ag(110) surface and   b) acetaldehyde following ethanol adsorption on the pre-oxidized Ag(110) surface. The coverages saturate the monolayer. Heating rate - 5 K/s.

Figure 4 compares the vibrational spectra of trifluoro-ethanol and trifluoro-ethoxide on the Ag(110) surface. As in the case of ethanol the molecular species is only weakly perturbed by adsorption exhibiting vibrational modes corresponding to all the modes observed in the liquid (*12*). These are all listed in Table II. Most notable are the modes due to the O-H group and both the symmetric and asymmetric stretching of the CF_3 group. The strong dipole strength of the CF bond makes these modes much more intense than CH stretch modes (*13*). The formation of the trifluoro-ethoxide by adsorption on the pre-oxidized surface results in a spectrum in which most of the vibrational modes are positioned at the same frequencies as in the molecular adsorbate with the obvious loss of the modes corresponding to vibration of the O-H bond. The second clear difference between the spectrum of the trifluoro-ethoxide and

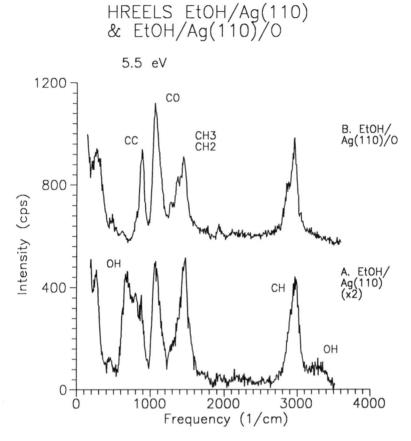

Figure 2. HREEL spectra of a) ethanol on the Ag(110) surface (I_{el} = 7.1 x 10^4 cps, FWHM = 60 cm^{-1}) and b) ethoxide formed by adsorption of ethanol on the pre-oxidized Ag(110) surface (I_{el} = 4.0 x 10^4 cps, FWHM = 65 cm^{-1}).

the trifluoro-ethanol is the disappearance of the asymmetric CF_3 stretch mode at 1260 cm^{-1}. This is attributed to a re-orientation of the molecule on the surface.

Acetic Acid and 2,2,2-trifluoro-acetic Acid. The surface chemistry of formic acid and acetic acid on both clean and pre-oxidized surfaces has been studied in some detail in the past (6,8). The desorption of acetic acid from the two surface is illustrated in Figure 5 and is the same as has been observed by other investigators. Acetic acid is adsorbed reversibly on the clean surface desorbing during heating at 195K with a heat of desorption of 11.5 kcal/mole. As in the case of ethanol, oxygen atoms on the surface serve as chemical bases to deprotonate the acid forming water and, in this case, an acetate group. The water desorbs from the surface at 200K leaving the surface covered with just

Table I. Vibrational Modes of Ethanol and Ethoxide on Ag(110)

Mode	Liquid (11)	Monolayer $C_2H_5OH/Ag(110)$	Ethoxide $C_2H_5O/Ag(110)$
ν_{Ag-O}			270 cm^{-1}
CH$_3$ twist	270 cm^{-1}	260 cm^{-1}	270 cm^{-1}
δ_{CCO}	427	440	475
γ_{OH}	674	680	
CH$_2$ rock	775	800	
ν_{CC}	877	850	890
ν_{CO}	1060	1080	1060
CH$_2$ wag		1300	1290
$\delta_{CH2}, \delta_{CH3}$	1408 1472	1380 1460	1370 1460
ν_{CH}	2834 2980	2980	2915
ν_{OH}	3360	~3300	

the acetate groups. During heating these then decompose resulting in CO_2, CH_3CO_2H, C_2H_2O and CH_4 desorption at 610 K and some deposition of carbon onto the surface.

The adsorption of 2,2,2-trifluoro-acetic acid on the surface results in chemistry that is much different from that observed for acetic acid. The desorption spectrum following adsorption of trifluoro-acetic acid on the clean Ag(110) surface is shown in Figure 5c and is identical to that observed following adsorption on the pre-oxidized surface. Trifluoro-acetic acid adsorbs irreversibly on both surfaces. The adsorbed species decomposes at 620 K resulting in desorption of CO_2 and fluorinated species, not all of which have been fully identified. Although the adsorbed species has not been identified spectroscopically, we assume at this point that it is 2,2,2-trifluoro-acetate which has been produced by deprotonation of the acid by the clean surface. This is consistent with both the observed products and the temperature at which it is decomposing on the surface.

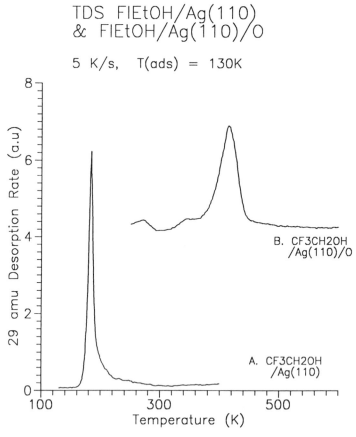

Figure 3. Desorption of a) 2,2,2-trifluoro-ethanol (CF_3CH_2OH) adsorbed on the clean Ag(110) surface and b) 2,2,2-trifluoro-acetaldehyde following CF_3CH_2OH adsorption on the pre-oxidized Ag(110) surface. Coverages saturate the monolayer. Heating rate - 5 K/s, m/e - 29 amu.

Discussion

Alkoxide Stabilization by Fluorination. The basic chemistry of the ethanol observed in this work is identical to that proposed from previous desorption experiments in that ethanol is deprotonated to form ethoxide on the pre-oxidized surface (5). The HREEL spectra shown here identify the surface species as those proposed. Other studies of the longer chain alcohols indicate that the same chemistry is observed for the series methanol-pentanol (9). The only systematic change as the alkyl chain is lengthened is that the heat of adsorption of the alcohols increases incrementally from 9.7 kcal/mole for methanol by ~1.1 kcal/mole/CH_2 group. The alkoxides formed by adsorption of

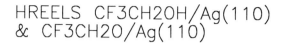

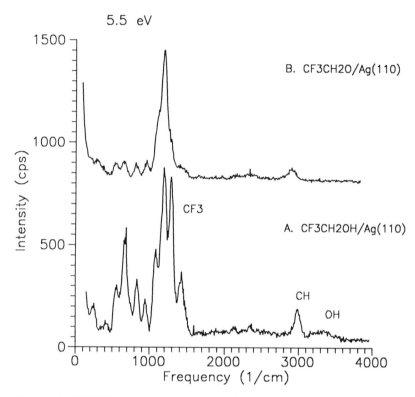

Figure 4. HREEL spectra of a) CF_3CH_2OH adsorbed on the clean Ag(110) surface (I_{el} = 5.1 x 10^4 cps, FWHM = 60 cm^{-1}) and b) 2,2,2-trifluoro-ethoxide produced by adsorption of CF_3CH_2O on the pre-oxidized Ag(110) surface (I_{el} = 2.2 x10^4 cps, FWHM = 60 cm^{-1}).

alcohols on the pre-oxidized surface all decompose over a fairly narrow temperature range around 280K to form aldehydes. The narrow temperature range indicates that the alkoxide decomposition kinetics are all identical and hence that the step initiating decomposition is the same in all cases. The fact that the primary product in all cases is the aldehyde indicates that the mechanism is one involving breaking of a C-H bond at the C_1 carbon atom (5).

There has been very little prior examination of the surface chemistry of fluorinated amphiphiles. The 2,2,2-trifluoro-ethanol used in this work exhibits two interesting features, not observed in the hydrocarbon alcohols, which are a direct result of fluorination of the alkyl chain. The basic chemistry is the same as that of ethanol in that the molecule adsorbs reversibly on the clean

Table II.   Infrared and HREELS bands for CF_3CH_2OH

Mode	IR - gas (12)	IR - Liq. (12)	Monolayer CF_3CH_2OH /Ag(110)	Ethoxide CF_3CH_2O /Ag(110)
δ_{CF3}	548 cm^{-1} 664	533 cm^{-1} 663	540 cm^{-1} 670	550 cm^{-1} 660
γ_{OH}		~700	670	
ν_{CC}	830	829	850	830
CH$_2$ rock δ_{OH}	940	945	940	960
ν_{CO}	1088	1080	1065	1040 (unres.)
ν_{CF3}^{s}	1183	1163 1145	1180	1195
ν_{CF3}^{a}	1263 1292	1247 1278	1265	1265
δ_{CH2}	1455	1460	1410	1445
ν_{CH}			2955	2905
ν_{OH}			~3300	

surface but is deprotonated by the pre-oxidized surface to generate the alkoxide. The heat of adsorption of the reversibly adsorbed fluoro-ethanol is 10.9 kcal/mole which is very similar to the number determined for ethanol itself (10.7 kcal/mole). The influence of fluorination manifests itself in the properties of the trifluoro-ethoxide. The first difference appears in comparing the vibrational spectra of the hydrocarbon versus fluorocarbon alcohols and alkoxides. In the case of the hydrocarbon alcohols it appears from quantitative XPS measurements that the alcohols are adsorbed with their alkyl chains lying parallel to the surface (9). This observation comes from XPS measurements of the relative concentrations of carbon and oxygen in monolayer films of the alcohols ethanol through pentanol. At saturation the relative concentrations of

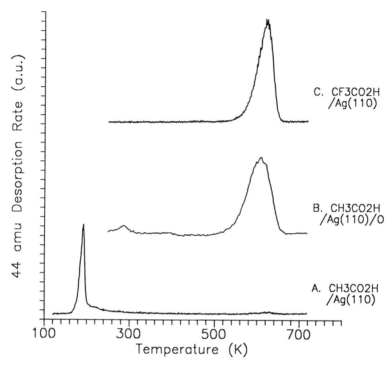

Figure 5. a) Desorption of acetic acid adsorbed on the clean Ag(110) surface. b) Desorption of CO_2 during thermal decomposition of acetate produced by adsorption of acetic acid on the pre-oxidized Ag(110) surface. c) Desorption of CO_2 following adsorption of 2,2,2-trifluoro-acetic acid on the clean Ag(110) surface. Heating rate - 5 K/s, m/e - 44 amu.

oxygen and carbon vary in a fashion consistent with the molecule being oriented with its alkyl chain parallel to the surface rather than perpendicular to the surface. The differences between the vibrational spectra of ethanol and ethoxide are primarily limited to the disappearance of the modes associated with the O-H group. Other than that there is a small increase in the intensity of the CO stretch mode. We have compared vibrational spectra of alkoxides and alcohols in set of alcohols methanol through pentanol (*14*). The conversion of methanol to methoxide results in a large increase in the intensity of the CO stretch mode. Although there is also an increase in the CO stretch intensity accompanying the conversion of the longer chain alcohols to their alkoxide it is much less than that observed for the methanol-methoxide conversion. We have interpreted this as indicating that the longer chain alkoxides do not undergo the same re-orientation as methoxide during formation and hence that they are oriented with their alkyl chains tilted away from the normal and interacting with the surface.

The HREEL spectra of the trifluoro-ethoxide are significantly different from those of trifluoro-ethanol. In the spectrum of trifluoro-ethanol both the symmetric (1180 cm^{-1}) and asymmetric (1265 cm^{-1}) stretch modes of the CF$_3$ group are clearly observed, giving very strong losses due to their large dynamic dipole moments. On formation of the trifluoro-ethoxide one observes the same features in the spectrum with the exception of the loss of modes associated with the O-H group and a disappearance of the asymmetric stretch mode of the CF$_3$ group. The loss of the O-H modes is simply the result of breakng of the O-H bond, however, the dissappearance of the asymmetric CF$_3$ stretch is the result of molecular re-orientation on the surface. If a particular mode is dipole active then the energy loss cross section will be dependent on the orientation of the molecule with respect to the surface (15). Modes with their dynamic dipoles normal to the surface will have high energy loss cross sections. Dipoles oriented parallel to the surface will induce opposing image dipoles in the metal and hence be inactive dipole scatterers since the adsorbate-metal system has no net dipole. Although trifluoro-ethanol is of low symmetry it has local C$_{3v}$ symmetry about the CF$_3$ group. If the C-C bond is oriented along the surface normal the symmetric CF$_3$ stretch will have a dipole moment along the surface normal while the asymmetric stretch will have a moment parallel to the surface. The change in the spectra of trifluoro-ethanol on forming the trifluoro-ethoxide is consistent with a structure in which the molecular adsorbate is initially oriented with its C-C bond tilted away from the surface normal. Dissociation of the O-H bond results in a structural transformation in which the C-C bond axis becomes aligned with the surface normal. In the case of the formation of ethoxide from ethanol no such reorientation is observed. Furthermore it should be noted that formation of the fluorinated propoxide and butoxides from the corresponding alcohols is not accompanied by such an obvious re-orientation (14).

The second obvious difference between the hydrocarbon and fluoro-carbon alcohols is in the stability of the ethoxides formed by their adsorption on the pre-oxidized Ag(110) surface. Whereas the hydrocarbon ethoxide decomposes during heating at 270K the trifluoro-ethoxide is stable to a temperature of 420K. This increased stability is also observed for the longer chain fluoro-alcohols and on Cu surfaces (14). The decomposition process appears to be the same for both fluoro-carbon and hydrocarbon alcohols in that the products are aldehydes and hence the mechanism involves dissociation of the C-H bond at the C$_1$ position. Fluorination inhibits dissociation and as such sheds some light on the nature of the transition state for this process. Fluorination of the methyl group results in a greatly increased electronegativity of this group. If the C-H bond breaks by hydride loss then the transition state is one in which the C$_1$ atom is cationic. Fluorination of the methyl group and increasing its electronegativity destabilizes the cation, increasing the energy of the transition state in the C-H bond breaking process and thus increases the kinetic barrier to dissociation. Our observation is consistent with a mechanism of hydride elimination as the initiating step in the alkoxide and fluoro-alkoxide decomposition process.

Carboxylic Acid Stabilization by Fluorination. The primary difference between the surface chemistry of the acetic and trifluoro-acetic acid is that while the former adsorbs reversibly on the clean Ag(110) surface the latter is adsorbed irreversibly. While acetic acid desorbs on heating at 195 K with a heat of adsorption of 11.5 kcal/mole the trifluoro-acetic acid is stabilized in the sense that it spontaneously deprotonates to produce the trifluoro-acetate which, rather than desorbing, decomposes on heating at 620K. One possible cause for this is that fluorination increases the heat of adsorption of the acetic acid to the point that rather than desorbing during heating it deprotonates. However, this seems unlikely given the fact that fluorination has a negligible effect on the heat of adsorption of ethanol. Instead, this chemistry can be explained with the same type of argument used to rationalize the stabilization of the trifluoro-ethoxide. If the adsorbed acids dissociate to form a carboxylate species by loss of a proton then the transition state is a carboxylate anion on the surface. Fluorination of the methyl group results in a stabilization of this anion and hence a lowering of the transition state energy and an increase in the deprotonation kinetics. In addition, fluorination of the methyl group probably stabilizes the final state as it does for the perfluorinated acids in aqueous solution where they are much stronger acids than their hydrocarbon analogues. Whereas acetic acid has a pK_a of 4.72, trifluoro-acetic acid has a pK_a of 0.23 (*16*). Although the data is not presented here the longer chain hydrocarbon acids and fluorinated acids exhibit the same behavior. While the hydrocarbon acids adsorb reversibly on the clean surface the perfluorinated acids are deprotonated at low temperature to yield the perfluoro-carboxylate. These then decompose on heating at higher temperatures (620 K).

Conclusions

The role of perfluorination in surface chemistry of potential boundary layer additives is to influence the kinetics of reactions in which the fluorinated alkyl chain can act to stabilize or destabilize ionic transition states. As in the case of alcohols on Ag(110) surfaces the fluorinated alcohols $CF_3(CF_2)_nCH_2OH$ adsorb reversibly on the clean surface but are deprotonated on the pre-oxidized surface to form the fluorinated alkoxides $CF_3(CF_2)_nCH_2O$ / Ag(110). The alkoxides decompose by dissociation of a C-H bond to form the aldehyde. The C-H bond dissociation is a hydride loss reaction involving a cationic transition state. The cationic C_1 atom is destabilized by the electronegative trifluoro-methyl group decreasing the decomposition kinetics and making the fluorinated alkoxide much more stable on the surface than the hydrocarbon alkoxide. The effects of fluorination in the carboxylic acids manifest themselves in much lower aqueous phase pK_a values resulting from stabilization of the carboxylate anion by the electronegative perfluoro-alkyl chain. On the Ag(110) surface stabilization of the carboxylate anion stabilizes the transition state involved in acid deprotonation and hence increases the kinetics of formation of the carboxylate species. Fluorination serves to stabilize the alkoxides formed on Ag(110) surface by decreasing the kinetics of the decomposition process. In contrast fluorination stabilizes the carboxylic acids on the clean Ag(110) surface by inducing the formation of the carboxylate at temperatures at which the hydrocarbon acids would desorb.

Acknowledgments

This work was supported by the Air Force Office of Scientific Research under grant no. AFOSR 89-0278. AJG holds a Fellowship in Science and Engineering from the David and Lucile Packard Foundation and is an A.P. Sloan Foundation research fellow.

Literature Cited

1. Bowden, F.P.; Tabor, D. *The Friction and Lubrication of Solids*, Oxford Press, 1950
2. Snyder, C.E.; Gschwender, L.J.; Tamborski, C. *Lubr. Eng.* **37**, *1981*, 344
3. Snyder, C.E.; Dolle, R.D., *ALSE Trans.* **19**, *1976*, 171
4. Walczak, M.M.; Leavit, D.K.; Theil, P., *J. Am. Chem. Soc.* **109**, *1987*, 5621
5. Wachs, I.E.; Madix, R.J., *Appl. Surf. Sci.* **1**, *1978*, 303
6. Barteau, M.A.; Madix, R.J., *Surf. Sci.* **120**, *1982*, 262
7. Wachs, I.E.; Madix, R.J., *Surf. Sci.* **76**, *1978*, 531
8. Sexton, B.; Madix, R.J., *Surf. Sci.* **105**, *1981*, 177
9. Zhang, R.; Gellman, A.J. in press *J. Phys. Chem.*
10. Parker, B.; Gellman, A.J. to be published
11. Krishnan, R., *Proc. Ind. Acad. Sci.* **A53**, *1961*, 151
12. Travert, J.; Lavalley, J.C., *Spectrochimica Acta* **32A**, *1976*, 637
13. Sverdlov, L.M.; Kovner, M.A.; Krainov, E.P., *Vibrational spectra of Polyatomic Molecules* J. Wiley, NY
14. Dai, Q.; Gellman, A.J. in press *Surf. Sci.*
15. Ibach, H.; Mills, D.L., *Electron Energy Loss Spectroscopy and Surface Vibrations*, Academic Press, *1982*, NY
16. McMurry, J. *Organic Chemistry*, **1984**, p. 746, Brooks/Cole, CA

RECEIVED October 21, 1991

Chapter 12

Tribochemical Reactions of Stearic Acid on Copper Surface

Infrared Microspectroscopy and Surface-Enhanced Raman Scattering

Zu-Shao Hu[1], Stephen M. Hsu[2], and Pu Sen Wang

Ceramics Division, National Institute of Standards and Technology, Gaithersburg, MD 20899

The tribochemistry of stearic acid on copper was studied using a pin-on-disc apparatus under boundary lubrication condition. The wear of copper was reduced significantly by the presence of the stearic acid. Fourier transformed infrared spectroscopy (FTIR) analysis suggests the formation of cupric stearate during the rubbing process. The reaction products were further confirmed to by Laser Raman spectroscopy. The characteristic Raman signals of cupric stearate at 1547, 623, 288, and 243 wavenumbers were detected. Static experiments at 160°C produced the same reaction products when compared to room temperature rubbing experiments.

The tribochemical reaction is postulated to be initiated by the formation of a chemisorbed stearic acid on copper oxide surface. The adsorbed specie appears to be a monomer and is oriented approximately perpendicular to the surface. The native oxide film is necessary for the chemisorption to occur. Interaction of the surfaces converts the chemisorbed stearic acid into cupric stearate which forms a protective boundary lubricating film reducing wear.

Stearic acid has been widely used as an antiwear agent in lubricants under boundary lubrication conditions (1). It is generally agreed that stearic acid functions by adsorption onto the rubbing surfaces (1-4). Evidence in the literature suggests that this adsorption is primarily due to adsorbed monolayers (2-3). Kramer (4) studied copper surface hardening rates as a function of stearic acid concentrations in paraffin oil and suggested the formation of copper stearate on the surface. Buckley (5) also indicated that a protective surface film might have formed during lubrication. However, there is no direct experimental evidence to substantiate the hypothesis.

---

[1]Visiting scientist from East China University of Chemical Engineering, People's Republic of China
[2]Corresponding author

Experimental

The stearic acid and hexadecane were purchased from Morton Thiokol Inc. The purities were 99% and the materials were used without further purification. The cupric stearate was manufactured by ICN Biomedicals Inc.

A copper rod made by Johnson Matthey Inc. of certified 99.9998% purity was machined into discs and pins. The surface was cleaned by Norton K624 crocus cloth, washed with acetone, etched in 2N hydrochloric acid for five minutes, and finally rinsed by deionized water and dried (27). The treatment was carried out in air at room temperature except for the samples to be studied under argon.

Stearic acid exists in three morphological forms: A, B, and C (28). The A form is rare and was not studied in this work. The B form was prepared by crystallization from the saturated solution in n-hexane at room temperature (29); and the C form was obtained by irreversibly heating the B form above 46°C (22).

For adsorption studies, stearic acid was adsorbed on the copper surface by dipping the copper into a 2.5×10^{-3} mol/liter stearic acid solution in n-hexane for approximately one hour [30]. Samples to be studied under argon atmosphere were prepared by cleaning and dipping in the solution and then sealing the specimens in a glass cell under argon.

For wear studies, the stearic acid was dissolved in hexadecane to form a 0.4 weight percent solution. The lubricant was applied to the copper disk surface to give an approximately $20\mu m$ film. The sample was used in a pin-on-disk wear tester under dry air at room temperature. The load was 15 newtons, the wear track diameter on the disk was 16 mm, and the linear sliding speed was 0.38 m/s. After 100 m or 400 m of accumulated sliding distance, the disk was analyzed. The tribochemical reaction products were extracted with tetrachloromethane for IR analyses.

The infrared spectroscopy study was conducted on a micro-FTIR system with a mercury cadmium telluride detector. The IR microscope was operated in reflection or transmission mode with a circular sampling area of $100\mu m$ in diameter. The resolution of the spectrometer was 4 cm^{-1}. The films on the copper surface were studied directly using the reflective mode. The pure copper stearate was pressed into a disc with KBr, and the isolated reaction products were coated on KBr. They were analyzed by the IR instrument in the transmission mode.

Raman spectra were obtained with the instrument in the microsampling mode in back-scattering configuration. The excitation of 488.0 nm wavelength was obtained using an argon ion laser. The incident laser power was varied from 1 to 10 mW. The microscope sampling area was $6\mu m$ in diameter. A triple monochromator was used to filter and disperse the Raman scattered signal, and an intensified diode array operating at -30°C was used to detect the frequency-dispersed signal from the triple monochromator. The spectral resolution was estimated to be about 4 cm^{-1}.

Results and Discussion

Wear tests were conducted with and without 0.4% by weight of stearic acid in hexadecane. Fig. 1 shows the micrographs of the worn surfaces and the associated surface profiles after wear. The wear track with stearic acid has a blue surface film (Fig. 1b). No film was observed for hexadecane alone (Fig. 1a). Note the wear depth was reduced from 1.35 μm to 0.32 μm by the stearic acid under the same test conditions.

Since stearic acid can morphologically exist in three different forms, different Raman scattering patterns may be expected. In our case, only B and C forms are relevant (28).

The Raman spectra of pure stearic acid in B and C crystal forms as well as that of the cupric stearate are shown in Fig. 2. Table I lists all the observed bands and their suggested assignments (24, 31-38). The symmetric stretch mode of the carbonyl group is responsible for the weak bands at 1636 cm^{-1} and 1649 cm^{-1} in the C and B forms of the stearic acid respectively. This ban shifts to 1547 cm^{-1} and becomes weaker in cupric stearate. The Raman peaks from the C-C' (bonded to the carboxyl) stretch at 909 cm^{-1} for the C form and at 912 cm^{-1} for the B form of stearic acid disappear in cupric stearate. The O=C-O in-plane deformation at 668 cm^{-1} in C form stearic acid shifts to 678 cm^{-1} in cupric stearate. The three bands at 623 cm^{-1}, 288 cm^{-1}, and 243 cm^{-1} were observed for the cupric stearate. These bands are characteristic of the Cu-O bond in the cupric stearate molecule. In the cupric stearate, no band at 1419 or 1407 cm^{-1} can be detected as in the stearic acid (39). The downshift of the skeletal vibration mode at 1104 cm^{-1} for the stearic acid to 1094 cm^{-1} for the cupric stearate and the disappearance of the longitudinal acoustic mode (LAM7) at 340 cm^{-1} indicate that they have different lengths in the carbon chain (23). The difference is attributed to the fact that stearic acid exists as dimers due to hydrogen bonding, while cupric stearate exists as monomers.

Fig. 3 represents the Raman spectra of stearic acid adsorbed on two different copper surfaces. The spectrum of the sample prepared and analyzed in an argon atmosphere is shown in Fig. 3a. This sample represents the stearic acid adsorbed on a copper surface without the oxide films. The bands associated with the carboxyl group remain unchanged at 1636 cm^{-1} and 909 cm^{-1} but the band at 668 cm^{-1} is too weak to be observed. Hence, the adsorption of stearic acid on copper surface without the oxide layer is probably physical adsorption. When the sample was prepared in air, a native oxide film was formed (40-41). As stearic acid adsorbs on the Cu_2O film, the band at 1636 cm^{-1} shifts to 1540 cm^{-1}, and the band at 909 cm^{-1} disappears as shown in Fig. 2b. In addition, these two spectra have neither 243 cm^{-1} nor 288 cm^{-1} bands which are characteristic of the cupric stearate. The band at 629 cm^{-1} (Fig. 3b) is at a higher frequency and shows more intensity than that in the cupric stearate spectrum. It must be due to the Raman scattering from a new species formed on the copper surface with Cu_2O. It appears to be different from both stearic acid (668 cm^{-1}) and cupric stearate (623 cm^{-1}) and is likely a chemisorbed complex of stearic acid on the surface. The surface oxide, Cu_2O, is therefore necessary for the formation of this complex.

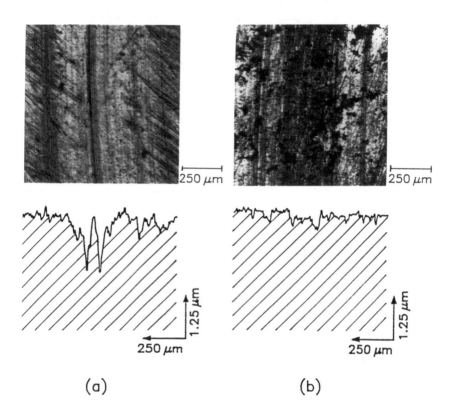

Fig. 1 Copper surface wear scars and profiles after pin-disc test when lubricated with (a) hexadecane, (b) hexadecane containing 0.4 wt.% stearic acid.

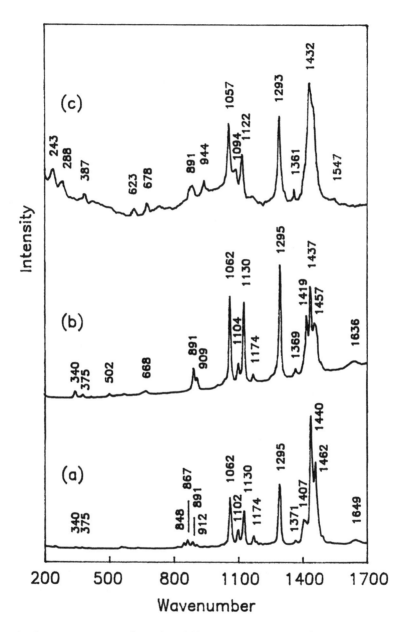

Fig. 2 Raman spectra of stearic acid in (a) B, (b) C crystal forms, and (c) cupric stearate.

Table I
Raman Results of Stearic Acid, Cupric Stearate,

Hexadecane, and Tribological Tests

			Copper Surface Lubricated by			
C Form Stearic Acid	B Form Stearic Acid	Cupric Stearate	Stearic Acid in Hexadecane after 100m	Stearic Acid in Hexadecane after 100m	Hexadecane after 400m	Assignment
1636	1649	1547			1540	C=O stretch
1457	1462					CH$_2$ rock
1437	1440	1437	1437	1437	1437	CH$_2$ bending
1419	1407					CH$_2$ bending
1369	1371	1361	1361	1361	1361	CH$_3$ sym. deform
1295	1295	1293	1300	1300	1290	CH$_2$ twist + CH$_2$ rock
1174	1174	1169				CH$_2$ twist + CH$_2$ rock
1130	1130	1122	1133	1128	1122	C-C stretch + C-C-C deform
1104	1102	1094	1081	1081		C-C stretch + C-C-C deform
1062	1062	1057	1060	1060	1057	C-C stretch + C-C-C deform
	946	944	957	957		C-H deform or CH$_2$ rock + CH$_2$ twist
909	912					C-C'(carboxyl) stretch
893	891	891	891	891	891	CH$_3$ rock + C-C stretch
	867					CH$_2$ rock + CH$_2$ twist
	848					CH$_2$ rock + CH$_2$ twist
668	678					O=C-O deform
		623	~615	629	623	Cu-O
					491	Cu-O in cupric oxide
			404			
340						LAM7
		288			288	Cu-O
		243			243	Cu-O
			217	214		

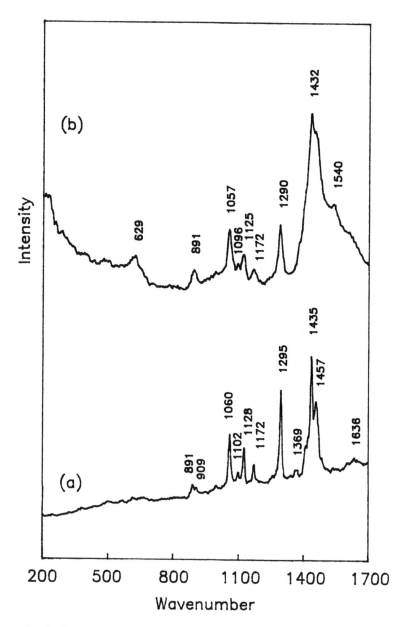

Fig. 3 Raman spectra of stearic acid adsorbed on copper surface treated (a) in Ar and (b) in air.

The interaction of the hydroxyl group in an alcohol solution with a copper surface suggests that the possible interaction between stearic acid and the native oxide film on a copper surface can be (42-43):

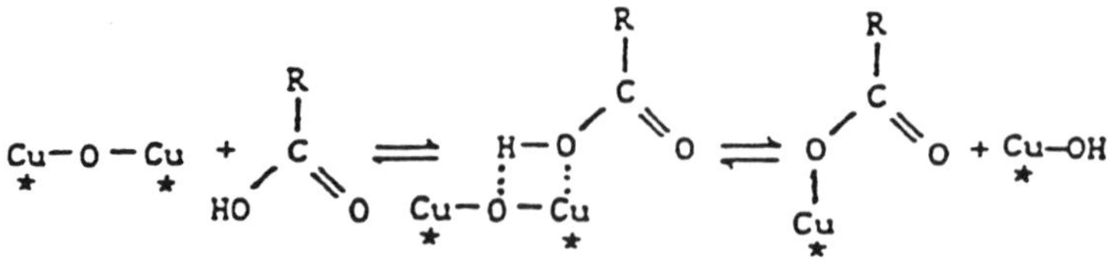

$$
\text{Cu}-\text{O}-\text{Cu} \ + \ \underset{\overset{|}{\text{HO}}}{\overset{R}{C}} \overset{\diagdown}{\underset{\diagup}{O}} \ \rightleftharpoons \ \text{H}-\text{O} \overset{\diagup}{\underset{\text{Cu}-\overset{..}{\text{O}}-\text{Cu}}{}} \overset{R}{\underset{|}{C}} \overset{\diagdown}{\underset{\diagup}{O}} \ \rightleftharpoons \ \text{O} \overset{\diagup}{\underset{\overset{|}{\text{Cu}}}{}} \overset{R}{\underset{|}{C}} \overset{\diagdown}{\underset{\diagup}{O}} \ + \ \text{Cu}-\text{OH} \qquad (1)
$$

(I)   (II)

The unidentate copper stearate complex (II) is bonded on the surface and can be transformed to bidentate (III) or bridging (IV).

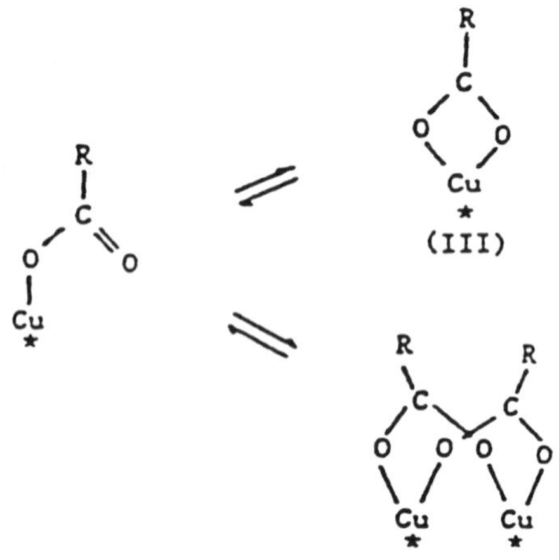

$$(2)$$

(III)

(IV)

In the case of the chemisorbed surface (Fig. 3b), the skeletal vibration frequency of 1102 cm^{-1} was down-shifted (Fig. 3a) to 1096 cm^{-1}. This indicates that the adsorbed complex exists on the copper surface in monomer form rather than as dimers. Since no signal of CH$_2$ bending at 1400 - 1420 cm^{-1} was detected, the adsorbed complex does not show the characteristic of any crystal structures (B or C). Moreover, the Raman scattering near 1460 cm^{-1} due to the Fermi resonance with the first overtone of the infrared-active CH$_2$ rocking modes (44-45) has not been observed when the stearic acid is adsorbed on the copper surface. This signal can normally be detected for stearic acid below its melting point (25). The SERS theory predicts a stronger enhancement for bands associated with molecules closer to the surface (46-47). Literature also indicates that SERS will preferentially enhance the vibration modes perpendicular to the metal surface (48). The relatively strong bands of Cu-O, C-C, and CH$_2$ suggest that the carbon chain of the chemisorbed stearic acid is most likely nearly perpendicular to the surface.

Fig. 4 shows Raman spectra of a copper surface (a) lubricated with hexadecane before rubbing, (b) after rubbing, (c) lubricated with hexadecane containing 0.4% stearic acid after rubbing for 100 meters sliding distance, and (d) after 400 meters sliding distance. Table 1 gives the observed frequencies and their assignments. Corresponding IR spectra are shown in Figs. 5 and 6.

When the copper surface was lubricated with pure hexadecane on the pin-on-disc test, the Raman scattering of the disc surface shows neither the chemisorbed hexadecane nor the reaction between the surface and hexadecane (Fig. 4a, 4b). The bands of surface copper oxide were too weak to be observed clearly compared to the strong bands of hexadecane especially in Fig. 4a. The OH group is active in both IR and Raman, but its stretching frequency is above 3000 cm^{-1} and is out of our spectral range. Fig. 4a and 4b mainly show the scattering modes from CH vibration from hexadecane covered but not chemisorbed on the disc surfaces.

The IR spectra of the surface lubricated with hexadecane are shown in Fig. 5. They were taken from the regions outside (Fig. 5a) and inside (Fig. 5b) of the contact area after the pin-disc experiment. The spectrum of the outside region represents the IR of pure hexadecane. The strong band from 2850 to 2965 cm^{-1} can be assigned to the symmetric and asymmetric vibrations of -CH$_2$- and -CH$_3$ groups, respectively. The 1461 cm^{-1} band is due to the scissoring of -CH$_2$- or asymmetric deformation of -CH$_3$. The peaks at 1372 and 1296 cm^{-1} are from the symmetric deformation of -CH$_3$ and the wagging of -CH$_2$-, respectively. And finally, the 717 cm^{-1} signal represents the vibrational energy of (-CH$_2$-)$_n$ when n is equal to or greater than four.

The broad band near 950 cm^{-1} on the contact point (Fig. 5b) is the main difference in IR spectra between the regions inside and outside the contact point. This band can be assigned to δ(OH) (-OH deformation) or ν(CO) (-CO- stretch) in alcohols, ν(OO) in peroxides, δ(OH) in acids, δ(=CH) in alkenes or δ(CH) in aldehydes. All these species are the possible reaction products of hexadecane after rubbing (49). However, no corresponding peaks associated with ν(C=O) and ν(C=C) were detected so that we can conclude the existence of acid, aldehyde, or alkene is unlikely. The reasonable assignment of this band is hence from peroxides or alcohol. This agrees with the weak absorption band near 3550 cm^{-1} due to the ν(OH) in alcohols or hydroperoxides.

After 100 meters of sliding with 0.4 wt. % stearic acid in hexadecane in the pin-on-disc wear tester, the spectrum (Fig. 4c) exhibits similar features to that of chemisorbed stearic acid (Fig. 3b). The strong band at 629 cm^{-1} indicates that the chemical interaction of stearic acid with copper surface has taken place and a chemisorbed complex was formed on the surface. We believe that this surface specie plays a role in protecting the surface from wear. The bidentate complex is the most stable form of the three because the carboxylate is in a polyligand and coordinates with the copper by two atoms. The mechanism of the formation of a protective film on the surfaces may be postulated as following:

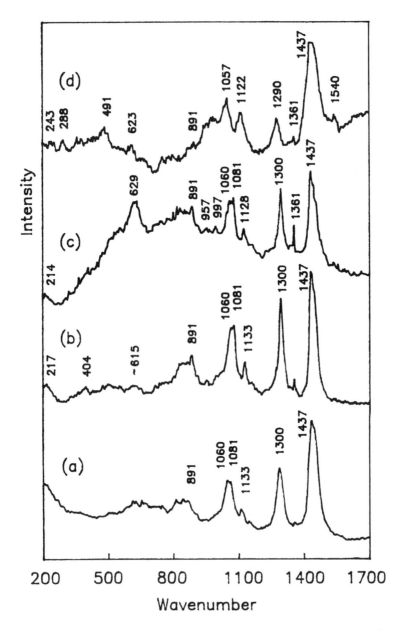

Fig. 4 Raman spectra of (a) hexadecane on copper surface before rubbing, (b) after rubbing 100 meters, (c) stearic acid in hexadecane on copper surface after rubbing 100 meters, and (d) after rubbing 400 meters.

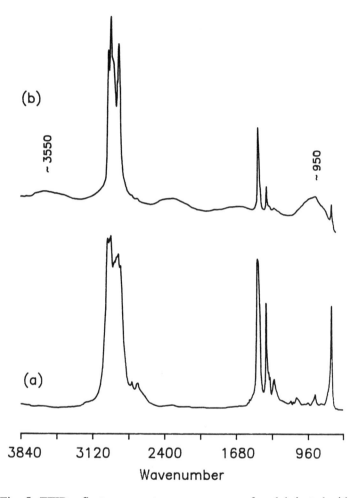

Fig. 5 FTIR reflectance spectra on a copper surface lubricated with hexadecane after pin-disc test: (a) outside, (b) inside of wear scar.

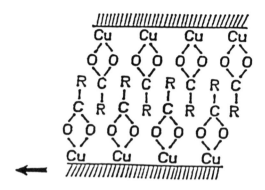

As sliding continues, the surface complexes can be separated due to wearing to form cupric stearate.

$$CuOOCR \xrightarrow{\quad} Cu(OOCR)_2 \tag{3}$$
$*$

Water and cupric oxide were produced by the reaction of surface hydroxyls

$$Cu\text{-}OH \xrightarrow{\quad} Cu=O + H_2O \tag{4}$$
$*$

Cupric oxide as shown in the 491 cm^{-1} band of Fig. 4d is the unique product of the tribochemical reaction. No cupric oxide was observed in a separate study on the thermal reaction up to 240°C (41). The intermolecular reaction of the chemisorbed complexes forms copper stearate. The Raman spectrum of the copper surface after sliding 400 meters (Fig. 4d) shows the scattering of carboxyl at 1540 cm^{-1}, CuO band at 623, 288, and 243 wavenumbers.

Fig. 6a shows the IR spectrum of copper stearate in hexadecane. Compared to the spectrum of pure hexadecane (Fig. 5a) three new bands (1585cm^{-1}, 1560cm^{-1}, and 1535cm^{-1}) were observed and they can be identified as frequencies in different structures of the carboxyl group in the copper stearate. The carboxyl group in a carboxylate salt, such as copper stearate, has multielectronic π bond ($\pi_{4/3}$). The coupling of the two C-O vibration modes gives separate symmetric and asymmetric stretch frequencies. The band at 1585cm^{-1} has been reported by Satake and Matuura (50) and assigned to structure I. Nakamoto (51) considered that the carboxylate ion may coordinate with a metal and form one of the structures (II to IV). The $\nu^a(CO_2^-)$ and $\nu^s(CO_2^-)$ of free carboxylate ion are ca. 1560 and 1416 cm^{-1}, respectively. In the unidentate complex (structure II), $\nu(C=O)$ is higher than $\nu^a(CO_2^-)$ and $\nu(C\text{-}O)$ is lower than $\nu^s(CO_2^-)$. The opposite trend is observed in the bidentate complex (structure III). In the bridging complex (structure IV, similar to structure I), however, two $\nu(CO)$ are close to the free ion values. Accordingly, the band at 1585 cm^{-1} can be assigned to $\nu^a(C=O)$ in structure II, the band at 1560 cm^{-1} to $\nu^a(CO)$ in free carboxylate ion (or structure IV or I), and the band at 1535 cm^{-1} to $\nu^a(CO)$ in structure III. Fig. 6c is the spectrum of copper surface coated with hexadecane containing 0.4% stearic acid. Note that the three peaks which are characteristics of

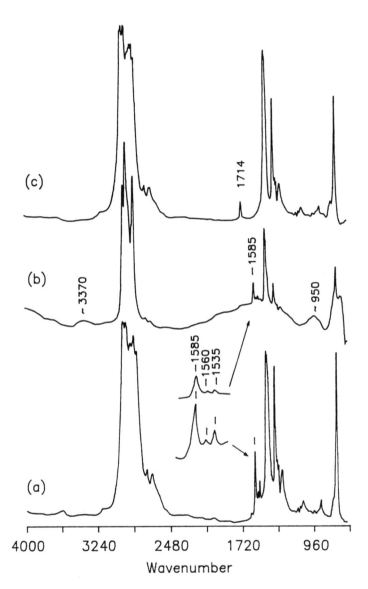

Fig. 6 FTIR (a) transmission spectrum of copper stearate in hexadecane, (b) reflectance spectra of copper surface lubricated with 0.4 wt.% stearic acid in hexadecane after pin-disc test, (c) before pin-disc test.

copper stearate are not detected. Instead, an IR signal at 1714 cm^{-1} was observed and this is characteristic of the carbonyl bond of stearic acid in hexadecane. The reaction of stearic acid with copper or copper oxides was not observed in this case from the spectrum. Fig. 6b represents the IR spectrum obtained from a copper surface rubbed with hexadecane containing 0.4% stearic acid. In addition to the two bands at 3370 and 950 cm^{-1} as we have identified to be peroxides or alcohol, the three peaks at 1585 cm^{-1}, 1560 cm^{-1}, and 1535 cm^{-1}, which are fingerprint of copper stearate in hexadecane were observed. This evidence strongly suggests the formation of copper stearate on the rubbed surface due to the tribochemical reaction of stearic acid and the copper surface. The 3370 cm^{-1} peak, which is 180 cm^{-1} lower in frequency than 3550 cm^{-1} observed previously, may suggest a stronger intermolecular hydrogen bond in the peroxides or alcohols resulted from the oxidation of hexadecane in presence of stearic acid.

Fig. 7 shows the transmission IR spectra of (a) solid copper stearate and (b) isolated reaction products from a copper surface rubbed with hexadecane containing 0.4% stearic acid. In Fig. 7a, the 1585 cm^{-1}, 1560 cm^{-1}, and 1535 cm^{-1} bands are the asymmetric stretch frequencies of carboxylate group as mentioned above. The 1443 cm^{-1}, 1415 cm^{-1}, and 1400 cm^{-1} bands are the symmetric stretch frequencies for bidentate, free carboxylate, and unidentate complex, respectively. In Fig. 7b, the 1714 cm^{-1} peak is the absorption of carbonyl group from the unreacted stearic acid. The peaks at 1585 cm^{-1}, 1560 cm^{-1}, 1535 cm^{-1}, and at 1400 cm^{-1} have exactly the same frequencies as shown in Fig. 7a. The band shapes are also similar though some difference can be observed. The assignment of bands at 1624 cm^{-1} in Fig. 7a and 1602 cm^{-1} in Fig. 7b are uncertain. Low et al. (14) suggested that the presence of oleic acid in copper oleate can shift the carboxylate band at 1585 cm^{-1} to a higher frequencies. It is possible that the frequency at 1585 cm^{-1} of copper stearate shifts to 1602 or 1624 cm^{-1} because of the presence of stearic acid as impurity, as suggested by Low et al.. Although the tribochemical reactions are complex and many products are formed, the IR spectra prove conclusively that copper reacts with the stearic acid to form copper stearate under boundary lubrication conditions.

Table II gives the relative intensities of asymmetric carboxylate frequencies normalized to the 1560 cm^{-1} band for copper stearate and isolated tribochemical reaction products shown in Fig. 7. In this table, the relative intensity of 1602 cm^{-1} band in the isolated reaction products is higher than that of 1624 cm^{-1} band in copper stearate. This is understandable because there is more stearic acid in the reaction products than in the copper stearate. If the suggestion by Low et al. applies, the intensity at 1602 cm^{-1} should be higher than that of the 1624 cm^{-1}. The copper stearate to reaction products intensity ratios at 1585 cm^{-1} band and 1535 cm^{-1} band are in opposite trend. Taking copper stearate as reference, the intensity at 1585 cm^{-1} in the reaction products decreases while at 1535 cm^{-1} increases. We believe this is wear-dependent. The 1585 cm^{-1} band which is contributed by unidentate complex is most likely to form during the beginning stage of rubbing when the adsorbed stearic acid reacts with the native copper oxide. As rubbing continues, the unidentate layer wears out and the reaction among stearic acid, adsorbed oxygen, and copper metallic surface produces the bidentate layer which is characteristics of 1535 cm^{-1} band (52). These spectra provide supporting evidence for our postulated tribochemical mechanism.

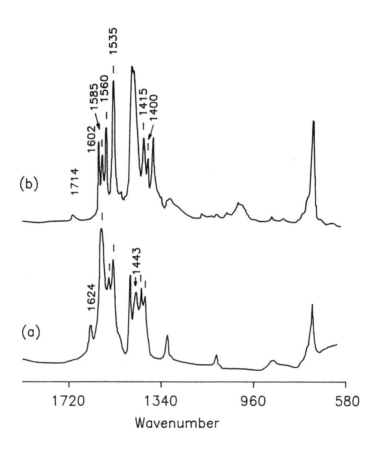

Fig. 7 Transmission infrared spectra of (a) solid copper stearate, (b) isolated tribochemical reaction products from a copper surface rubbed with hexadecane containing 0.4 wt.% stearic acid.

Table II

Relative Intensities of Carboxylate Asymmetric Frequencies
(normalized to the intensity at 1560 cm^{-1})

Sample	I_{1585}	$I_{1624\ or\ 1602}$	I_{1535}
Copper stearate (Fig. 4a)	1.64	0.43	1.21
Reaction products (Fig. 4b)	0.70	1.09	1.48

Next, we want to examine the nature of the tribochemical reaction. The infrared spectra of hexadecane containing stearic acid on copper after thermal treatment at various temperatures are shown in Fig. 8. After heating at 65°C for five minutes, the carbonyl absorption band of stearic acid in hexadecane shifts from 1714 cm^{-1} (Fig. 6c) to 1698 cm^{-1}. This coincides with solid stearic acid. In addition, bands at 1530 cm^{-1} and 1390 cm^{-1} were detected. They are due to the chemisorbed stearic acid. Their frequencies are near but lower than the asymmetric and symmetric frequencies of C-O group in bidentate structure of copper stearate. No such bands can be found in stearic acid or copper stearate separately. This proves that the chemisorbed species have the similar structure to bidentate copper stearate and the bond between the carbonyl group with copper surface is stronger than that in bidentate. A similar observation was reported by Low et al. for oleic acid on copper surface (14). The ester-like species are bonded to the surface according to structure V. But in our case, the surface species are more likely to resemble structure I according to the absorption at 1530 cm^{-1} and 1390 cm^{-1}. The hydroxyl group forms a hydrogen bond with the carbonyl group of the adsorbed stearic acid and causes the frequency shift of the carbonyl group. Carbonyl frequencies in structure V are at 1585 cm^{-1} (asymmetric) and at 1400 cm^{-1} (symmetric) while in structure I they are possibly at 1530 cm^{-1} (asymmetric) and at 1390 cm^{-1} (symmetric).

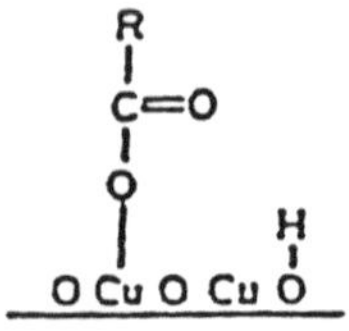

$$
\begin{array}{c}
R \\
| \\
C=O \\
| \\
O \\
| \qquad\qquad H \\
| \qquad\qquad | \\
\underline{\quad O\ Cu\ O\ Cu\ O\quad}
\end{array}
$$

V

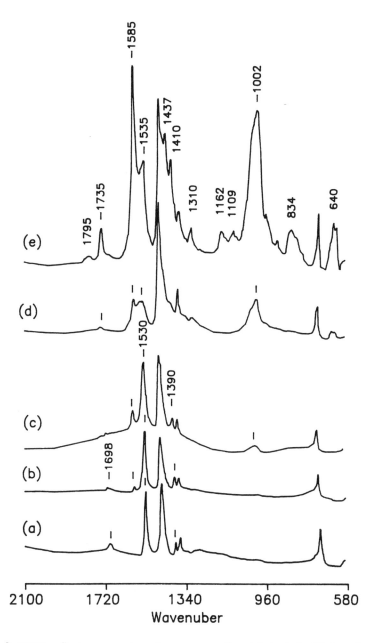

Fig. 8 FTIR reflectance spectra of a copper surface lubricated with hexadecane containing stearic acid after heating five minutes at (a) 65°C, (b) 110°C, (c) 125°C, (d) 140°C, and (e) 160°C.

This lack of direct evidence is probably due to the complexity of the chemical system and the limited methodology capable of detecting and characterizing the minute amount of the tribochemical reaction products.

Infrared spectroscopy (IR) has been used extensively for hydrocarbon analysis. It is surface sensitive and is often adapted for surface analysis (6-8). Micro-Fourier transformed infrared (FTIR) spectroscopy has recently been successfully demonstrated to study the interaction between tricresyl phosphate in paraffin oil with steel surfaces during rubbing (9).

Several attempts to elucidate the tribochemical reaction mechanism between stearic acid and metal surfaces were made (10-11). Eberhardt and Mehliss reported their preliminary work using IR on stearic acid absorbed on an oxidized copper surface (12). The absorption bands at 1585, 1560, and 1520 cm^{-1} for copper organometallic bond were not identified. Fatty acids on steel surfaces were characterized by infrared emission spectroscopy at 100, 150, and 200°C (13), and the formation of iron soaps was suggested. Again there was no definitive spectrum assignment due to the complexity of the surface chemistry.

The reaction of oleic acid with copper surface was also studied by IR emission and reflection spectroscopy (14). Results suggest that the formation of an ester-like species. The spectra were tentatively identified as a copper bi-ester, a precursor of a bi-nuclear copper complex. Kamata et al. used attenuated total reflectance (ATR) spectroscopy to study the stearic acid Langmuir-Blodgett monolayers on evaporated silver films (15). Jakosen studied stearic acid adsorbed on iron surface from 1% solution in hexadecane using ATR and observed the iron stearate bands (16).

Raman scattering is less sensitive to perturbations from neighboring molecules. The Raman spectra in the low frequency region, which are characteristic of the vibrational modes of many surface complexes formed by the reaction between lubricants and the surfaces, can be more readily interpreted. Surface-enhanced Raman scattering (SERS) provides an opportunity to study the surface reactions. The SERS signals from the absorbed species may provide direct information on the bonding between the surface and the adsorbate. Examples abound in literature for the in-situ probing of corrosion inhibiting complexes on copper surfaces, organic sulfide absorbed on silver, etc. (17-20). However, in tribology, fluorescence interference arising from oxidized hydrocarbons and the surface damages from the wear process make the application of micro-Raman spectroscopy to tribochemistry difficult. There is no prior work reported on the use of Raman spectroscopy for the study of the stearic acid/copper system under boundary lubrication conditions. Some work has been reported on the determination of crystalline structures of stearic acid (21-26).

We have succeeded in obtaining both the micro-FTIR and micro-Raman spectra on copper surface. This paper describes the results of the interactions between copper and stearic acid in hexadecane solution, under boundary lubrication conditions.

At 110°C, the 1585 cm^{-1} band appears (Fig. 8b). This band gains its intensity as temperature increases. Starting from 125°C, two bands at 1002 cm^{-1} and 1735 cm^{-1} were detected. They are characteristics of esters resulting from the oxidation reactions. The spectrum at 140°C (Fig. 8d) is very similar to Fig. 6b in the frequency range of carboxyl and carbonyl groups. At 160°C (Fig. 8e), the carboxylate bands and ester bands become stronger. Thus, the tribochemical reaction products under our conditions are similar to the thermochemical reaction products at the temperature range of 140-160°C.

Conclusion

Surface enhanced Raman scattering has shown that cupric stearate differs from stearic acid mainly in the Cu-O bands at 623, 288, 243 cm^{-1} and the shift of C=O band from 1640 cm^{-1} to 1547 cm^{-1}. The C-C' (carboxyl) band of the stearic acid at about 910 cm^{-1} is also undetectable in cupric stearate. The native cuprous oxide on a copper surface is necessary for the stearic acid chemisorption. Raman study of the tribochemical reaction between hexadecane containing 0.4 wt% of stearic acid with the oxidized copper surface suggests the formation of the chemisorbed complex on the surface by bonding a carboxyl group to the surface. We postulate that the orientation of this chemisorbed stearic acid molecule is approximately perpendicular to the surface and that it is in monomer form. This chemisorbed species is responsible for the protection of the copper surface. Further rubbing of the surface will convert this species to cupric stearate. Comparing IR results of the tribochemical and thermochemical reactions, we conclude that the tribochemical process in the pin-disc experiment is basically comparable to the thermochemical reaction of stearic acid with a copper surface at 140°C to 160°C temperature range. In both cases, three asymmetric bands of carboxylate in the copper stearate can be observed. However, the esters are formed only in the thermochemical reaction but not in the tribochemical process. The chemisorbed stearic acid (structure I) appears only in the thermochemical reaction below 140°C and not in the pin-on-disc experiments. The antiwear species in the protective film formed during the tribochemical reactions are most likely chemisorbed complex acid or bidentate copper stearate.

Acknowledgement

The financial support from the Energy Conversion and Utilization Technology (ECUT) tribology program, Department of Energy is sincerely appreciated.

Certain commercial equipment, instruments, or materials are identified in this paper in order to adequately specify the experimental procedure. Such identification does not imply recommendation or endorsement by the National Institute of Standards and Technology, nor does it imply that the materials or equipment identified are necessarily the best available for the purpose.

Literature Cited

1. Wills, J. G., Lubrication Fundamentals, Marcel Dekker, New York, (1980), 33.
2. O'Connor, J. J. and Boyd, T., Lubrication Engineering, McGraw-Hill, New York (1968), Chapter 2.
3. Ling. F. F., Klaus, E. E., and Fein, R. S., Boundary Lubrication and Appraisal of World Literature, American Society of Mechanical Engineers, New York (1969), 87.
4. Kramer, I. R., "The Effect of Surface-Active Agents on the Mechanical Behavior of Aluminum Single Crystals," Trans., AIME, 221, 989, (1961).
5. Buckley, D. H., Surface Effect in Adhesion, Friction, Wear and Lubrication, Elsevier, New York, (1981), 542.
6. Boerio, F. J. and Chen, S. L., "Infrared Spectra of Absorbed Films on Metal Mirrors," J. Colloid and Surf. Sci., 73 (1), 176, (1980).
7. Allara, D. L. and Swalen, J. D., "An Infrared Reflection Spectroscopy Study of Oriental Cadmium Arachidate Monolayer Films on Evaporated Silver," J. Phys. Chem., 86 (14), 2700, (1982).
8. Golden, W. G., Snyder, C. D., and Smith, B., "Infrared Relection-Absorption Spectra of Ordered and Disordered Arachidate Monolayers on Aluminum," J. Phys. Chem., 86, 4675, (1982).
9. Hegemann, B. E., Jahanmir, S., and Hsu, S. M., "Microspectroscopy Applications in Tribology," Microbeam Analysis, 193, (1988).
10. Daniel, S. G., "The Adsorption on Metal Surfaces of Long Chain Polar Compounds From Hydrocarbon Solutions," Trans. Faraday Soc., 47, 1345, (1951).
11. Beischer, D. E., "Radioactive Monolayers: A New Approach to Surface Research," J. Phys. Chem., 57, 134, (1953).
12. Eberhardt, E. and Mehliss, G., "Eine Küvette zur Ultrarotspek Troskopischen Untersuchung der Chemisorption von höheren Fettsäuren an Aufgedampften Metallfilmen," Z. Chem., 248, (1961).
13. Low, M. J. D. and Inone, H., "Infrared Emission Spectra of Fatty Acids on Steel Surfaces," Canada J. Chem., 43, 2047, (1965).
14. Low, M. J. D., Brown, K. H., and Inone, H., "The Reaction of Cleic Acid With Copper Surfaces," J. Colloid and Interface Sci., 24, 252, (1967).
15. Kamata, T., Kato, A., Umenura, J., and Takenaka, T., "Intensity Enhancement of Infrared Attenuated Total Reflection Spectra of Stearic Acid Langmiur-Blodgett Monolayers With Evaporated Silver Island Films," Langimur, 3161, 1150, (1987).
16. Jakobsen, R. J., "Application of FT-IR to Surface Studies," Fourier Transform Infrared Spectroscopy Applications to Chemical System, 2, Academic press, 165 (1979).
17. Gardiner, D. J. Gorvin, A. C., Gutteridge, C., Jackson, A. R. W., and Raper, E. S., "In-Situ Characterization of Corrosion Inhibition Complexes on Copper Surfaces Using Raman Microscopy," Corrosion Sci., 25, 1019, (1985).
18. Jeziorowski, H. and Moser, B., "Raman Spectroscopic Studies of the Interaction of Oxalic Acid and Sodium Oxalate Used as Corrosion Inhibitors with Copper," Chem. Phys. Lett., 120 (1), 41, (1985).

19. Sandroff, C. J. and Herschbach, D. R., "Surface-Enhanced Raman Study of Organic Sulfides Adsorbed on Silver: Facile Cleavage of S-S and C-S Bonds," J. Phys. Chem., 86, 3277, (1982).
20. Allen, C. S. and Patterson, M. L., "Chemical Accessibility of SERS Active Structures on Anodized and Electroplated Copper Surfaces," J. Electroanal. Chem., 197, 373, (1986).
21. Warren, C. H. and Hooper, D. L., "Chain Length Determination of Fatty Acids by Raman Spectroscopy," Can. J. Chem., 51, 3901, (1973).
22. Vergoten, G. and Fleury, G., "Overall and Lattice Vibrations of Fatty Acids. 1 - C Form of Stearic Acid," J. Raman Spectros., 12(2), 206, (1982).
23. Rabolt, J. F., "Hydrogen Bonding Effects on the Skeletal Optical and the Longitudinal Acoustical Modes in Long Chain Molecules and Polymers," J. Polym. Sci.: Polym. Phys. Ed., 17, 1457, (1979).
24. Rabolt, J. F., "Frequency and Intensity Patterns of LAM Progression in Weakly Coupled Chains: The Case of Stearic Acid and Stearyl Alcohol," J. Chem. Phys., 81 (11), 4782, (1984).
25. Zerbi, G., Conti, G., and Minoni, G., Pison, S., and Bigotto, A., "Premelting Phenomena in Fatty Acid: An Infrared and Raman Study," J. Phys. Chem., 91, 2386, (1987).
26. Vergoten, G. and Fleury, G., "Transition Dipole-Dipole Coupling Interactions in B Form of Stearic Acid Single Crystals," Chem. Phys. Letters, 112 (3), 272, (1984).
27. Tanner, D. W., Pope, D., Potter, C. J., and West, D., "The Promotion of Dropwise Condensation by Monolayers of Radioactive Fatty Acids I. Stearic Acid on Copper Surfaces," J. Appl. Chem. (London), 14, 361, (1964).
28. Vergoten, G., Fleury, G., and Moschetto, Y., Advances in Infrared and Raman Spectroscopy, 4, Clark, R. J. H. and Hester, R. E., Heyden, London, (1978), Chapter 5.
29. Morishita, H., Ishioka, T., Kobayashi, M., and Sato, K., "Study of Micopolytype Structure in Crystals of Stearic Acid B Form by the Raman Microprobe Technique," J. Phys. Chem., 91, 2273, (1987).
30. Miller, S. K., Baiker, A., Meier, M., and Wokaun, A., "Surface-Enhanced Raman Scattering and the Preparation of Copper Substrates for Catalysis Studies," J. Chem. Soc. Faraday Trans. 1, 80, 1305, (1984).
31. Nanba, T. and Martin, T. P., "Raman Scattering From Metal Smokes," Phys. Stat. Sol. (a), 76, 235, (1983).
32. Mo, Y., Li, X., and Zhang, P., "SERS of the Molecules Adsorbed on Cu Surface at 4880Å Excitation," Chin. Phys. Lett., 2 (9), 413, (1985).
33. Taylor, J. C. W. and Weichman, F. L., "Raman Effect in Cuprous Oxide Compared With Infrared Absorption," Can. J. Phys. 49, 601, (1971).
34. Huang, K., "The Long Wave Modes of the Cu_2O Lattice," Z. Phys., 171, 213, (1963).
35. Carabatos, C., Diffine, M., and Sieskind, M., "Fundamental Vibrational Bands of the Cuprite Lattice," J. Phys. 29, 529, (1968).

36. Krantz, M., Rosen, H. J., Macfarlane, R. M., Lee, W. Y., Lee, V. Y., and Savoy, R., "Raman Microprobe Study of a Superconducting (Tc=120K) $Ti_2Ca_2Ba_2Cu_3O_{10}$ Thin Film," Solid State Comm. 69 (3), 209, (1989).
37. Mathey, Y., Greig, D. R., and Shriver, D. F., "Variable-Temperature Raman and Infrared Spectra of the Copper Acetate Dimer Cu_2 $(O_2CC_3)_4$ $(H_2O)_2$ and Its Derivatives," Inorg. Chem., 21, 3409 (1982).
38. Koyama, Y. and Ikeda, K, "Raman Spectra and Conformations of the cis-Unsaturated Fatty-Acid Chains," Chem. Phys. Lipids, 26, 149, (1980).
39. Strobl, G. R. and Hagedorn, W., "Raman Spectroscopic Method for Determining the Crystallinity of Polyethylene," J. Polym. Sci.: Polym. Phys. Ed., 16, 1181, (1978).
40. Poling, G. W., "Infrared Reflection Studies of the Oxidation of Copper and Iron," J. Electrochem. Soc. Solid State Science, 116 (7), 958, (1969).
41. Kubaschewski, O. and Hopkins, B. E., "Oxidation of Metals and Alloys," Butterworths, (1962).
42. Campbell, I. M., Catallysis at Surfaces, Chapman and Hall, New York, (1988), 143.
43. Bartok, M., Sterochemistry of Heterogeneous Metal Catalysis, John Wiley & Sons, New York, (1985), 297.
44. Abbate, S., Zerbi, G., and Wunder, S. L., "Fermi Resonances and Vibrational Spectra of Crystalline and Amorphous Polymethylene Chains," J. Phys Chem., 86, 3140,(1982).
45. Abbate, S., Wunder, S. L., and Zerbi, G., "Conformation Dependence of Fermi Resonances in n-Alkanes. Raman Spectra of 1,1,1,4,4,4-Hexadeuteriobutane," J. Phys. Chem., 88, 593, (1984).
46. Furtak, T. E. and Reyes, J., "A Critical Analysis of Theoretical Models for the Giant Raman Effect from Adsorbed Molecules," Surf. Sci., 93, 351, (1980).
47. Murray, C. A. and Allara, D. L., "Measurement of the Molecule-Silver Separation Dependence of Surface Enhanced Raman Scattering in Multilayered Structures," J. Chem. Phys., 76, 1290, (1982).
48. King, F. W., Duyne, R. P. V., and Schatz, G. C., "Theory of Raman Scattering by Molecules Adsorbed on Electrode Surface," J. Chem. Phys., 69, 4472, (1978).
49. Coates, J. P. and Setti, L. C., "Infrared Spectroscopy as a Tool for Monitoring Oil Degradation," ASTM Special Tech. Publication, 918 (Aspects Lubr. Oxid.), 57, (1986).
50. Satake, I. and Matuura, R., "Studies With Copper (II) Soaps, Part I. Structure Invesigations of Copper Soaps and Their Complexes With Pyridine and Dioxane in Solid State," Kolloid-Z, 176, 31, (1961).
51. Nakamoto, K., Infrared and Raman Spectra of Inorganic and Coordination Compounds, John Wiley & Sons, New York, p 232, (1987).
52. Kiselev, V. F. and Krglov, O. V., Adsorption and Catalysis on Transition Metals and Their Oxides, New York, 346, (1989).

RECEIVED October 21, 1991

Chapter 13

Surface Lubricity of Organic Films for Rigid Containers and Ends

P. J. Palackdharry and B. A. Perkett

Packaging Products Division, Dexter Corporation, One East Water Street, Waukegan, IL 60085

Experiments were conducted in our laboratory to characterize certain selected natural and synthetic waxes as additives to an organic film composition to promote surface lubricity. From this work, Carnauba or lanolin exhibited quite acceptable properties as an internal lubricant in polyvinyl chloride based coating systems.

Beer, beverage, and food products packed in metal containers are protected by organic coatings to maintain product integrity and original flavor. Coatings on the outside of containers and lids (ends) protect the metal from corrosion and provide aesthetic enhancement to the package. The coating, whether on the inside or outside of a can, must remain intact during the fabrication, handling, packing, and food storage processes. Coatings for rigid containers and can ends are fabricated from precoated aluminum or steel stock. The coatings must have adequate surface lubricity during the intricate can forming and end conversion processes. The surface lubricity is provided in part by synthetic and natural waxes incorporated into the film forming components. This paper will report on the effect of certain selected waxes on the lubricity of the film used for the fabrication of container ends from precoated coils and their respective influence on maintaining film integrity.

Background

Metal containers with the corresponding ends comprise an 87 billion can per year market in the U.S. Soup, dog food, meat, beer, soda, fruit, and vegetables are all examples of products which are packaged in cans. Specific amounts and type of waxes are incorporated in some

0097–6156/92/0485–0217$06.00/0

of the coatings as the fabrication and packed-product protection need dictate.

Electrostatic spray or curtain-coating methods are the most common application techniques used commercially to coat cans. When a coating is cured on a pre-formed can, there is no need for the coating to be subjected to the abrasion, bending, and stretching associated with fabrication. Thus, inside spray coating formulations for pre-formed cans contain little or no wax additive. An example is the spray coating on the interior of an aluminum soda can. Other types of inside spray coatings for steel food cans may contain a small amount of wax to protect the film during beading of the can or to aid in product release (from the inside wall of the can).

Roller-coating and electrocoating processes are used to apply a coating to flat metal sheets. Depending on the film weight applied, rigid containers could be drawn out of or ends fabricated from these coated sheets. The wax selection is critical to the success of converting the flat stock into the finished shape. If the interior coating is fractured, the product packed will be exposed to the metal and be contaminated. During fabrication, the coating and metal could be stretched as much as 300% while in intimate contact with the dies. Good surface lubricity of the dies will ensure that localized coating strain and coating pick-off are minimized or completely eliminated.

Container Ends

The focus of our work was to develop appropriate coatings from which can ends could be readily fabricated. An "end" (or lid) is the top portion of an aluminum beer or beverage can. Coatings containing internal lubes are applied on an aluminum coil by roll coating, cured in a gas-fired oven, and cooled with a water quench. A thin layer of wax is applied before the coated coils are wound. At the fabrication plant, disks are punched out and shaped into ends, each with a score area and a tab attached by a rivet. Each end, depending upon its size, is used to "cap" a specific sized can to hold the designated packed product.

End-coated aluminum stock has two waxes: an internal wax which is incorporated into the coating and a surface wax which is applied after the coating is cured. One purpose of the surface wax is to lower the coefficient of friction (COF) to provide good mobility of the coated stock through the high speed fabrication equipment. Obviously, an end that gets stuck in one of the dies will cause equipment downtime and possible damage to the tooling. The surface wax may also be preventing localized strain in the coating and metal substrate, which would result in less damage to the film, thus a better quality product.

Since the waxes used for these purposes are not film forming polymers, their concentration within the film matrix must be limited to maintain the required film integrity. This concentraion may range from 0-4% by weight depending on the surface properties of the film forming polymer together with application, cure, and fabrication conditions.

End Coating Compositions

In the U.S., polyvinyl chloride (PVC) based polymers are most commonly used to coat the inside of ends for beverage cans. Both dispersion (relatively high molecular weight) and solution type vinyls are commercially available. Some solution type PVCs (10,000-30,000 molecular weight) are modified with carboxyl or hydroxyl functionalities to promote adhesion and to participate in crosslinking reactions. Such behavior is absolutely necessary in order for these coating compositions to adhere to treated aluminum coils. Typical solution vinyl and dispersion vinyl coating formulations are shown in Table I.

Table I.   Coatings Composition

Raw Material	Wt % On Solids	Raw Material	Wt % On Solids
Solution Vinyl	70 - 100	Dispersion Vinyl	40 - 80
Epoxy	0 - 20	Solution Vinyl	20 - 60
Crosslinker	0 - 10	Crosslinker	0 - 10
Internal Lube	0 - 4	Internal Lube	0 - 4

Dry film thickness on the inside of the ends for beverage containers averages about 7 microns with the surface lubricant averaging about 0.1 micron. The coefficient of friction (COF) as measured on the Altek 9505A Mobility/Lubricity tester generally averages 0.01.

Waxes Used In Food Coatings

A wax may be described as an organic mixture or compound which is a solid at room temperature, has a drop point above 40 C, a low viscosity above its melt point, and is water repellent and odor-free. Waxes usually consist of a hydrocarbon or an alkyl acid, alcohol, or ester having a molecular weight of 5,000 or less. An oil or lubricant, while being similar to a wax, is generally con- sidered to be a liquid at room temperature. These terms are used loosely in this paper because some waxes, such as petrolatum, have both solid and liquid components.

The interior coatings on beer, beverage, or food cans come in direct contact with the packed product and therefore must comply with FDA Section 21 CFR 175.300. This regulation deals with resinous and polymeric coatings which may be safely used as food contact surfaces in producing, manufacturing, packing, processing, preparing, treating, packaging, transporting, or holding of food products intended for consumers consumption. The most common waxes used in these coatings include paraffin, petrolatum, lanolin, mutton tallow,

carnauba, polyethylene, PTFE, and silicones.   Table II lists
properties of several wax types.

The factors involved in choosing the best wax(es) for a particular
application are coefficient of friction, abrasion resistance,
recoatability, bloom, product resistance, non-stick/product release,
appearance, and ease of fabrication.

Another very important consideration is the effect of the wax on the
food being packaged, such as an off-flavor, color development, or
change of product appearance.

Coating formulations for can ends contain between 0.2 to 4.0 weight
percent wax in the dry film.   Silicones and other surface active
agents could be used effectively at even lower levels but are not
recommended for this particular application.   Wax type and level need
to be balanced with the coating's ability to withstand fabrication
and subsequent resistance to the packed product.

Experimental

A series of waxes were separately substituted and tested in the end
coating formulations shown in Table I.   The height of metal
deformation at fracture was measured using a Tinius Olsen Number
88250 Ductility Tester.   Surface lubricity (COF) was determined using
an Altek model 9505A Mobility/Lubricity tester.   A Wilkins-Anderson
Company model 10778 Digital Enamel Rater was used to measure metal
exposure after ends were fabricated.

The coatings were applied and cured on chrome-treated 0.0115 gauge
5182 alloy aluminum panels according to prevailing industrial
practice (8-16 seconds total oven time, peak metal temperature
400-475 F.)   After a water quench, panels were allowed to cool to
room temperature before any testing was performed.

The Tinius Olson Ductility Tester is one technique used to establish
how well the cured film is able to withstand the fabrication process.   A 3/4" diameter polished steel ball is slowly raised through a
fixed test panel.   Either the fracture pressure or fracture height
can be measured at the metal fracture point.

The Altek Mobility/Lubricity tester pulls a 2000-gram sled across a
horizontal test panel at 20 cm/minute.   The sled is supported by
three one-half inch diameter polished balls.   Stretching of a spring
is a measure of the force needed to move the sled across the panel.
The results are the dimensionless COF.

The digital enamel rater was used to test the porosity of a converted
end.   A panel is converted into an end in a four step process.   These
involve formation of the basic shell, rivet reversal, scoring of the
metal, and attaching the rivet.   The dies used in our laboratory
presses are similar to those used commercially.   The converted end is
then placed in contact with a 1.0 weight percent aqueous conductive

TABLE II: WAX TYPES FOR FOOD CAN COATINGS

Natural

Wax	Source	Melting Point (C)	Composition	
Lanolin	Sheep Wool	36 - 42	C_9-C_{31}	Acid Esters 55%
				Alcohol Esters 15%
				Sterols, Terpenes 30%
Carnauba	Palm Leaves	82 - 85	$C_{20}-C_{50}$	Acid Esters 85%
				Acid, Alcohols, Resins 15%

Mineral/Petroleum

Wax	Source	Melting Point (C)	Composition	
Paraffin	Crude Oil	45 - 70	$C_{18}-C_{36}$	n-alkane 50-90%
				iso-,cyclo-alkane 50-10%
Micro-crystalline	Crude Oil	60 - 100	$C_{36}-C_{60}$	n-alkane 50-95%
				iso-,cyclo-alkane 50- 5%
Montan	Lignite Coal	81 - 84	$C_{24}-C_{60}$	Acids, Free and Esterified 50-82%
				Ketones, Hydrocarbons, Esterified Alcohols 4-13%
				Resins, Asphaltenes 16-39%

Synthetic

Wax	Source	Melting Point (C)	Structure	
Polyethylene	Fossil/Ethylene	90 - 130	$C_{70}-C_{350}$	Linear to Branched
Fisher-Tropsch	$CO + H_2$	90 - 100	$\sim C_{50}$	Linear Only
Amide	Fatty Acid + Amine	60 - 140	$C_{18}-C_{40}$	Amide or Diamide
PTFE	N.A.	300+	$C_{100,000}$	Polytetrafluoroethylene
Silicones	SiO_2, C, CH_3Cl, H_2O	<25	SiO_{60-100}	Modified Polysiloxanes —

salt solution under a 6.3 VDC potential for four seconds. After this
time the milliamp reading is reported as the enamel rater value (ERV)
or the relative porosity of the organic film. The porosity is
proportional to the exposed metal caused by fracture of the organic
film.

Results and Discussion

As mentioned throughout this paper, internal lubricants are
incorporated in organic coatings to provide surface lubricity for the
efficient manufacture of metal cans and ends from coated stock. The
integrity of the film must be maintained during all fabrication
processes in order to protect the product to be packed. In addition,
the coating must remain intact over the recommended shelf life of the
respective packed product without contributing any extractable
component which might adversely affect flavor or color.

Additives. The work conducted in our laboratory to achieve these
performance criteria led to the evaluation of a wide series of
organic and inorganic additives. One additional objective was to
determine how these additives affected end fabrication when no post
(external) lubrication was applied prior to article fabrication.
This post lubrication step involves the application of 10 mg./sq. ft.
of petrolatum, lanolin/paraffin, or paraffin onto the cured organic
film.

Table III lists some of the pertinent materials which were incorporated in a standard organic coating composition and evaluated for
non-post lubrication capabilities.

Further work was done to combine materials which exhibited a positive
effect as an internal lubricant but were not altogether acceptable
for some specific property. For example, a low melting point amide
wax looked very promising at not requiring post lubrication of the
film, but the visual appearance was not acceptable because of
haziness and surface topography. However, a combination of this
lubricant with lanolin and/or petrolatum would provide a smooth
transparent film without any significant adverse effects on surface
lubricity or flexibility.

Correlate COF With Film Integrity. A series of experiments was
performed to determine whether any correlation could be drawn
between surface lubricity (COF) and film integrity during the
fabrication of container ends.

   Solution Vinyl Coating. In one experiment, 1% by weight of
different internal lubricants were incorporated into a solution vinyl
coating formulation and tested for COF and metal exposure.   No
surface wax was applied. These results are shown in Table IV.

Table III: Coating Additives

Internal Lubricant	Effect*
C_{38} Diamide	o
C_{22} Erucamide	+
C_{20} Behenamide	+
C_{18} Oleamide	+
Polyethylene (MW = 3000)	−
Highly Branched Polyethylene	−
Low Mol. Wt. Polyethylene	+
Polyethylene Glycol	+
Polytetrafluoroethylene (PTFE)	−
Silica	−
MoS_2	−
$CaCO_3$	−
Graphite	−
Silicones	−
Carnauba Wax	+
Lanolin	+
Microcrystalline Wax	+
Petrolatum	+
Fischer-Tropsch Wax	o

* Effect = Cured film COF and appearance together
          with flexibility on fabrication.

Key: − = Negative Effect (Overall)
       o = No Significant Change Versus Standard
       + = Positive Effect

Table IV.   Waxes in Solution Vinyl Coating

Wax Type	Approximate Particle Size (Microns)	Coating Appearance	COF (X4)	Metal Exposure of Converted Ends (mA)
Carnauba	3	Good	.039	9.2
Lanolin	N.A.	Good	.090	13
Petrolatum	N.A.	Good	.025	14
Polyethylene A	6	Good	.080	29
Polyethylene B	5	Fair-Good	.130	23
Polyethylene B	13	Fair-Good	.130	32
Polyethylene C	30	Poor	.150	37

The particle size of a particular wax type influenced its performance properties. This and similar experiments have shown that the coefficient of friction and metal exposure were lowest when 2-8 micron wax was incorporated in the coating, then increased rapidly as the larger wax particle size exposed more of the unprotected coating surface.

Epoxy/Phenolic Coating. In Table V, an epoxy/phenolic end coating containing 1.0 weight percent wax was evaluated without the use of surface wax.

Table V.   Waxes in Epoxy/Phenolic Coating

Wax	COF (x4)	Metal Exposure of Converted Ends (mA)
Carnauba	0.045	90
Lanolin	0.090	121
Fischer-Tropsch	0.090	127
PE	0.130	136
Silicone	0.140	> 200
Amide	0.400	> 200

The trend was that higher COF did cause greater fracture of the organic film during the elongative and compressive operations of article fabrication. Note that the silicone and amide used as internal lubricants cause much greater fracture of the resulting film. These product types are not to be used by themselves, but have been used in combination with other waxes in which synergistic effects promoted acceptable properties.

Developmental Coating. Similarly, a developmental coating incorporating several internal waxes was evaluated (Table VI). Unlike the dispersed PVC and epoxy/phenolic coatings, the developmental coating demonstrated a linear correlation between the COF and ERV (Figure 1) and was noticeably more incompatible with the waxes (hazy coating appearance and more wax bloom to surface).

Table VI.   Waxes in Developmental Coating

Wax	COF (x4)	Metal Exposure of Converted Ends (mA)
Commercial Blend	0.051	2.9
Carnauba	0.050	3.1
Lanolin	0.045	3.2
Fisher-Tropsh	0.150	5.9
Polyethylene	0.320	7.4

Stripping of Surface Wax. One method developed and tested in our laboratory was to establish the extent of film fracture which is masked by the surface lubricant. The test involved extracting the petrolatum surface lube from the fabricated end with an organic solvent. By comparing the enamel rater value (ERV) of stripped ends with unstripped ends, the true extent of metal exposure is found. Table VII lists an example.

Table VII.   Enamel Rater Values After Removal of Surface Wax

Internal Wax	Enamel Rater Value (ma) Before Strip	After Strip
Low Mol. Wt. PE	1.0	1.5
Lanolin	1.5	2.0
Microcrystalline - B	2.8	12.4
Microcrystalline - A	16.0	> 200

The first two lube combinations are quite acceptable for use in this organosol (PVC) formulation. Microcrystalline wax B illustrates how well a wax can prevent wetting of the substrate and mask the actual metal exposure. Typical enamel rater values increase 50-200% after the surface wax is removed. Removal of the surface wax improves the ability of the ERV solution to wet the surface.

Optimum Wax Level. Wax levels in dispersion vinyl coatings were varied from 0.1 to 4.0 weight percent. The optimum wax level was found to be between 0.5 to 1.0% wax (Figure 2). Wax levels over 1.0% tended to weaken the film matrix without significantly lowering the COF.

Actual wax level in a coating may need to be altered to satisfy other coating requirements.

Rivet Head Diameter. The rivet head eccentricity was determined on several converted can ends containing 0.2, 0.4, 0.8, and 1.6 weight percent wax. This was accomplished by using a vernier caliper to measure the rivet head diameter at three distinct positions on three different ends. The eccentricity is defined as the average of these nine measurements. It was hoped that the effect of fabricating an end at different COF's would be manifested in the shape of the center

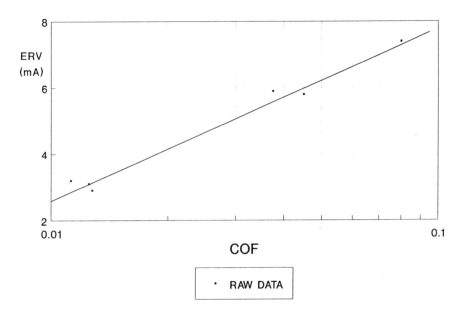

FIGURE 1.   ENAMEL RATER VALUE (mA) OF CONVERTED ENDS VERSUS
            COEFFICIENT OF FRICTION.

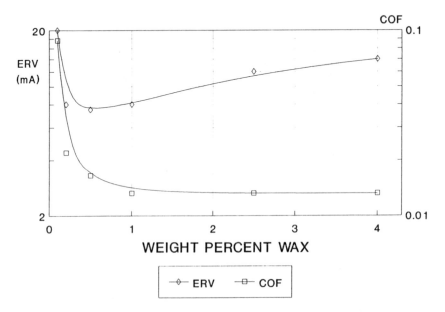

FIGURE 2.   ENAMEL RATER VALUE (mA) OF CONVERTED ENDS VERSUS
            WEIGHT PERCENT WAX IN CURED FILM.

Diameter (0.0001")   COF

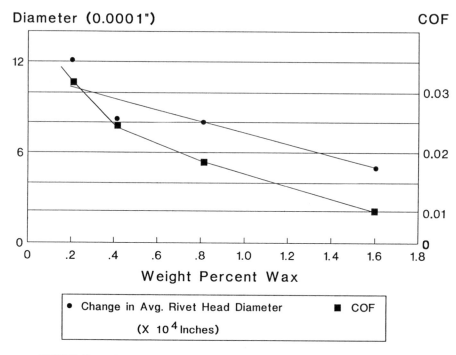

Weight Percent Wax

- ● Change in Avg. Rivet Head Diameter   ■ COF
  $(X\ 10^{4}$ Inches)

FIGURE 3.   CHANGE IN RIVET HEAD DIAMETER AND COEFFICIENT OF
FRICTION VERSUS WEIGHT PERCENT WAX IN CURED FILM.

rivet on the can end. Significant diameter eccentricity was found
only at the highest COF/lowest wax level (Figure 3).

Wax Volatility. Wax volatility was indirectly measured by the change
in the coating COF by using two different cure schedules. Two
identical experimental coatings were baked to a 450 F peak metal
temperature using total oven times of 13 and 18 seconds. It was
apparent that the longer cure allowed the more volatile oil to
vaporize, which increased the COF (Figure 4). Some waxes with high
flash points had COF's which were reduced slightly during the longer
cure. One explanation is that certain waxes require more time in
which to migrate to the air/coating interface during the thermal
curing cycle of the applied film.

Fracture Height. Metal fracture height, as measured by the Tinius
Olsen Ductility Tester, correlated well with the coating COF as long
as the wax particles were small enough to effectively cover the

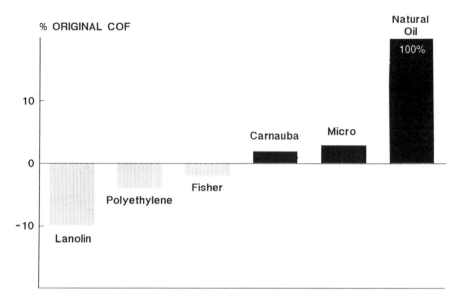

FIGURE 4.   PERCENT COEFFICIENT OF FRICTION CHANGE FROM SHORT
            BAKE SCHEDULE.

entire coating surface. The height at fracture, or strain at
fracture, was found to correlate well with the coating COF and to be
independent of the wax type (Figure 5). The variablity of the
individual fracture height measurements made many repetitions
necessary. This method is not recommended for comparitive studies.

Conclusion

Natural and/or synthetic waxes are incorporated in organic coatings
to provide adequate surface lubricity. These waxes, which are not
film formers, must not adversely affect the integrity of the organic
film while adding to the abrasion resistance and easy release of the
packed product. Use of Carnauba or lanolin wax with an average
particle size below 8 microns provided practically all the
performance requirements at a total concentration level of about 1%
in the film. Coated metal stock are subjected to a post external
lubrication step prior to article fabrication. None of the waxes and
combinations evaluated as internal film lubricants were totally
capable of eliminating the post lubrication step.

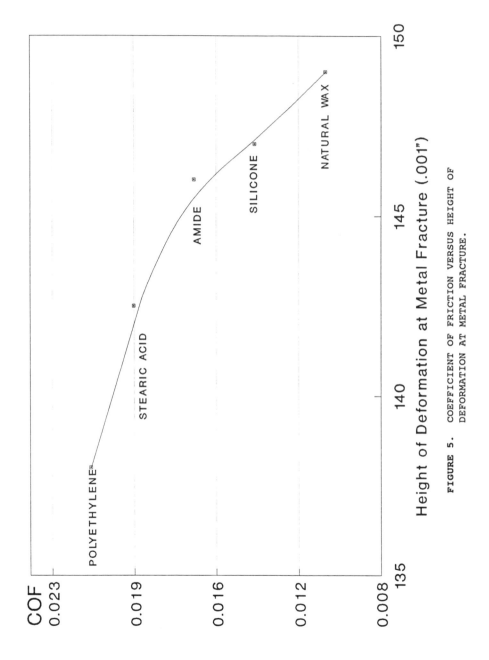

Height of Deformation at Metal Fracture (.001")

FIGURE 5. COEFFICIENT OF FRICTION VERSUS HEIGHT OF DEFORMATION AT METAL FRACTURE.

References

1. Industrial Waxes, Bennett, H., Vol. 1, Chemical Publishing
    Company, New York, **1963**
2. Unmuth, G.E., "Petroleum Waxes - Their Composition and Physical
    Properties," Presented at CSMA Convention, Chicago, **1975**
3. Handbook of Coatings Additives, Calbo, L.J., Ed., Page 271-280,
    Marcel Dekker, Inc., New York, **1987**
4. Jansen, K.H.M., Paint and Resin, **1988**, 58(2), 19
5. Frantz, M.C., American Paint and Coatings Journal, **1987**, 58
6. Kirk-Othmer Encyclopedia of Chemical Technology, Vol. 24, 3rd
    Edition, Pages 466-481, John Wiley & Sons, New York, **1978**
7. Michelman, J.S., and Homoelle, J.B., TAPPI Proceedings, 1988
    Polymer, Laminations, and Coatings Conference, **1988**
8. Fink, Ferdinand et.al., JOCT, **1990**, 62(791), 47
9. Frantz, M.C. and Pineiro, R., Daniel Products Company
10. Auger, C.J., JOCCA, **1989**, 9, 297
11. Wilson, W.R.D., Malkani, H.G., and Saha, P.K., "Boundary
    Friction Measurements Using A New Sheet Metal Forming
    Simulator", Department of Mechanical Engineering, Northwestern
    University, Evanston, IL, **1990**.

RECEIVED November 25, 1991

Chapter 14

Kinetics of Damage Generation on Single-Crystal Silicon Surfaces

The Influence of Lubricant Adsorption

Yuheng Li[1], Steven Danyluk[1], Selda Gunsel[2], and Frances Lockwood[2]

[1]Department of Civil Engineering, Mechanics, and Metallurgy, University of Illinois at Chicago, Chicago, IL 60680
[2]Pennzoil Products Company, The Woodlands, TX 77387

Single crystal silicon surfaces ((111) n-type with resistivity of approximately 30-Ωcm) were scratched at room temperature by a dead-loaded spherical diamond indenter (0.49-1.96 N) that was translated in the [110] direction at a speed of 5cm/sec. The scratches were placed between four electrical pads through which the resistivity of the damage region can be recorded. The surface of the silicon was covered with a commercial lubricant additive and the concentration of the additive and the load on the diamond were systematically varied. The change in voltage as a function of time was dependent on the concentration of the lubricant additive. This suggested that the adsorption and electrolytic properties of the lubricant additive are related to the extent of the subsurface damage produced by the diamond.

The chemisorption of lubricants on silicon surfaces has a good deal of influence on the mechanical processing of silicon into complex shapes. For example, mechanical damage can occur a result of abrasion during semiconductor processing, such as sawing, lapping, dicing and polishing. This mechanical processing is usually done in the presence of a lubricant that is supposed to reduce blade vibration and aid in removing the frictional heat. Lubricants can however react chemically with silicon surfaces and it has been found that the chemical reactions at silicon surfaces influence the mechanical damage generation and propagation [1]-[3].

There have been a number of experimental methods, such as etching, x-ray topography and electron microscopy to

0097–6156/92/0485–0231$06.00/0

determine the amount of the damage generated due to the mechanical processing of silicon [4]-[6]. These studies have shown that the damage consists of dislocations and microcracks, and the density of these defects depends on doping level, crystallographic orientation and the environment. There is not yet a quantitative theory to explain or predict the influence of the semiconductor properties and environments on the type or extent of damage.

Danyluk et. al.[7] developed an electrical resistivity measurement technique to quantify the mechanical damage produced in linear scratches and dicing grooves of instrumented spherical and Vickers diamonds and diamond-impregnated wheels. The technique involves fabricating a specially designed printed circuit on a silicon wafer. The circuit simulates a four point probe measurement that is used in the semiconductor industry to measure sheet resistance of wafers. The circuit can be used to measure changes in resistivity of the silicon as a dead-loaded diamond indenter scratches the silicon surface between the probes.

The present paper describes a real-time measurement of the change in resistivity which is related to the kinetics of the damage generation. Lubricant additives on the surface of the silicon modify the kinetic behavior. We report the results of a dynamic measurement of the change in resistivity during the scratching of single crystal silicon as a function of lubricating environmental conditions and dead loads. An adsorption model of deformation is proposed.

EXPERIMENTAL PROCEDURE

Single crystal silicon wafers (50 mm diameter, 0.33 mm thick, (111) n-type), supplied by the Monsanto Electronic Materials Company, were used for this study. The wafer surfaces are polished on one side and lapped on the other side. The resistivity was approximately 30 Ω-cm.

The wafers were processed as described previously [8] in order to produce a simulation of a four point probe measurement. Figure 1 shows a schematic diagram of the circuit. A constant DC current is supplied at the two outside pads and the change of the voltage was measured at the inside pads as a function of time for various scratching conditions and environmental variables. The experiment involves dead-loading a spherical diamond (nominal radius of 0.33 mm) and setting the diamond in motion at a speed of 50 mm/sec in the [110] direction. The diamond produces a single scratch that is positioned midway between the two inner metallization pads. The experiments are carried out in a clean room environment where the relative humidity and temperature are 20% and 78° F respectively.

De-ionized water and a commercial lubricant were placed on the silicon surface prior to the scratch formation, and these results were compared to the case with the surface free of lubricants. The commercial lubricant was an alkylated surfactant dissolved in de-ionized water to produce concentrations of 0.015, 0.029, 0.053, 0.076 and 0.100 wt%. The conductivities of these solutions ranged from 9 to 150 μmho/cm and was measured by Cole-Parmer Model 1484-10 conductivity meter. Figure 2 shows the results of conductivity versus the wt% of the lubricant additive.

The voltage-time measurements are recorded by a computer system at a sampling rate of 100,000 Hz per channel. The voltage resolution of the A/D converter is 0.01 mV. A schematic diagram of the system is shown in Figure 3.

Figure 4 shows a schematic diagram of the ball-on-flat geometry and the anticipated change in resistivity as a function of position. The compressive stresses ahead of the diamond and the tensile stresses behind the diamond produce damage so that the resistivity increases as a function of time as the diamond approches the probe positions, and a steady state change is produced after the diamond has passed the probe positions.

RESULTS

Figure 5 shows examples of CRT displays of the voltage versus time for these different surface environments (air, DI water and a lubricant at a concentration of 0.100 wt%) and a dead-load of 0.98 N (100 gf). In each case, there is a rise in voltage when the diamond passes by the metallization pads, and the increase in voltage relative to when there is no damage. The slope of the voltage versus time varies with environmental conditions. For example, the relative changes in voltage are 3.42, 6.17 and 3.75 % and with slopes of 106 mv/sec, 93 mv/sec and 70 mv/sec for the unlubricated surface, de-ionized and 0.100 wt% lubricant additive respectively. Other lubricant additive concentrations gave similar results.

The relative voltage change (%) versus time (ms) for all the lubricant additive concentrations at a dead-load of 0.49 N (50 gf) is shown in Figure 6. The change in voltage shows an increase of 3.27 % in air environment and 6.07 % in DI water. As the lubricant additive concentration is increased from 0 to 0.100 wt%, the voltage varies from 6.07 % to 2.45 %. The slope of the voltage-time plots for each of the lubricant additive concentrations varied and Figure 7 shows the slope of the voltage change as a function of lubricant additive concentration. The data show that the slope decreases from 97.4 to 50.8 mv/sec as the lubricant additive concentration increases from 0 to 0.100 wt%.

Figure 8 shows the relative change in voltage (%) as a function of the lubricant additive concentration (wt%) and load (N). The figure shows that as the load increases, the relative change in voltage also increases but the effect of lubricant is smaller.

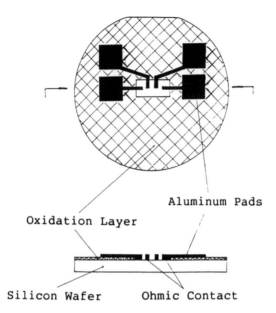

Aluminum Pads

Oxidation Layer

Silicon Wafer   Ohmic Contact

Figure 1. Schematic Diagram of the Sample for Dynamic Resistivity Measurement.

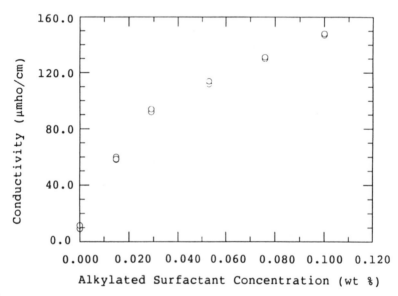

Figure 2. The Conductivity of the Solution as a Function of Additive Concentration.

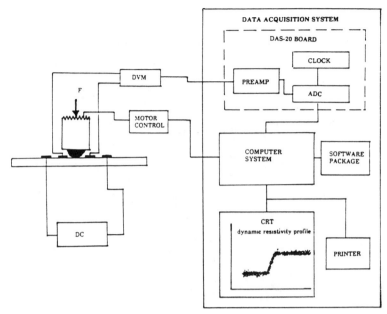

Figure 3. Schematic Diagram of the Scratching Apparatus and Data Acquisition System.

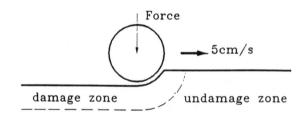

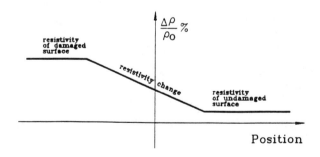

Figure 4. Schematic Diagram of the Ball-on-Flat Geometry and Expected Change in Resistivity as a Function of Time.

A

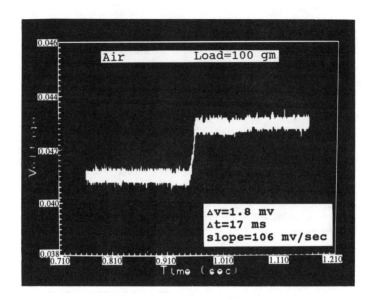

B

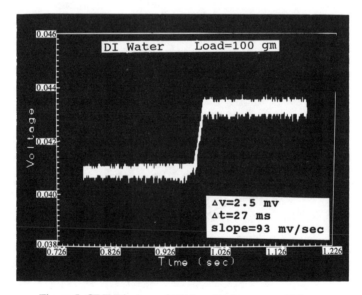

Figure 5. CRT Displays of Surface Environments Effect
on Dynamic Voltage. *Continued on next page.*

C

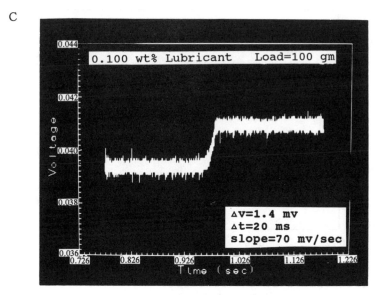

Figure 5. Continued.

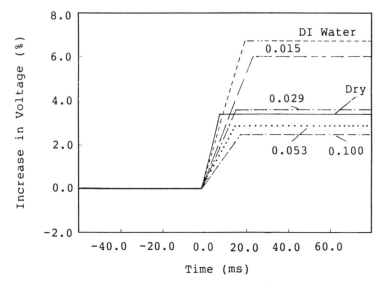

Figure 6. The Relative Change in Voltage versus Time.

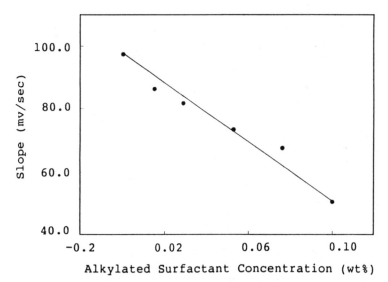

Figure 7. The Slope of the Voltage Change versus Alkylated
Surfactant Concentration.

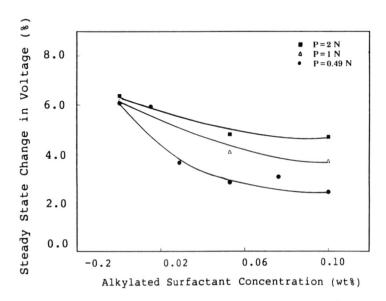

Figure 8. The Relative Change in Voltage versus Alkylated
Surfactant Concentration.

DISCUSSION

It is well known that the scratching damage in silicon is composed of microcracks and dislocations [9]-[10]. Dislocations are formed as a result of compressive stresses and microcracks are formed as a result of tensile stresses [11]. Dislocations and microcracks scatter electrons so that resistivity increases as the number of microcracks and dislocations increases [12]-[14]. This means that electrical resistivity is related to the amount of mechanical damage.

Environmental effects are also known to have an influence on the tribological behavior of semiconducting materials and a number of researchers have investigated the adsorption effects on surface mechanical properties [15]-[16]. It has been proposed that adsorption of gases and fluids either pin dislocation cores to the surface or modify the surface energy so that crack propagation is favorable. There is also some evidence that space charge effects can influence the surface deformation if the externally imposed loads generate stresses that have a spatial extent of the same size as the space charge region [17]. Surface charges affect the dislocation motion and damage formation and generation [18]-[19] because the size and magnitude of the surface charge determines the space charge fields. The magnitude and extent of the space charge depends on the surface charge, bulk doping and temperature, and for silicon at room temperature it can range from 0.5 to 10 μm below the surface [20]. These depths of the space charge are comparable to that of the yield stress surfaces for spherical indenters contacting silicon. Lubricating environments can modify the surface charge and consequently the deformation mode.

The experimental results presented in this paper show that the voltage versus time is correlated with the movement of the diamond and exhibits a time dependent and steady state value. The former depends on the adsorption kinetics and damage propagation as the stresses vary, and the latter is related to the total damage. The model for the steady state change in voltage $(V-V_o/V_o)$ has been previously examined and found to be related to the damage size and the conductivity of the damage zone for a linear scratch made by a diamond indenter [7]:

$$\frac{V-V_o}{V_o} \simeq \frac{b^2}{d^2} \left[1 - \frac{2\sigma_2}{\sigma_1} \right] \tag{1}$$

where V and V_o are the measured voltage with and without damage, b is the subsurface damage cross sectional radius, d is the inner probe spacing, and σ_2 and σ_1 are the conductivities of the subsurface damage zone and the silicon substrate, respectively. Equation (1) expresses the voltage in terms of measured quantities and it is expected that voltage will change as the damage size increases or the conductivity decreases.

The effective size of the damage zone, b, can be modified by adsorption if the yield surface is within the space charge fields. The space charge fields in semiconductors in contact with electrolytic fluids is described by a Debye length, which has been found to vary as [21]:

$$x = \frac{2\varepsilon_1}{\varepsilon_2} \left(\frac{\varepsilon_2 kT}{8\pi qN_d}\right)^{\frac{1}{2}} \left(\frac{V_s}{U_o + V_e}\right) \tag{2}$$

where ε_1 and ε_2 are the dielectric constants of semiconductor and electrolyte, q and N_d are the electron charge and doping density, V_s, V_e and U_o are the surface potential of the semiconductor, the applied potential across the diffusion layer of electrolyte and space charge potential at the interface and k and T are Boltzmann's constant and temperature respectively.

The above equation suggests that the thickness of the space charge region varies with ε_2 the dielectric constant of the electrolyte, temperature and doping level of the semiconductor. As a result, mechanical damage will depend on the lubricant additive concentration because the dielectric constant will vary with concentration.

Adsorption will influence damage if $b \approx x$. In this case:

$$\frac{V - V_o}{V_o} \approx \left\{ \frac{x^2}{d^2} \left[1 - \frac{2\sigma_2}{\sigma_1}\right]\right\} \tag{3}$$

so that the resistivity of the damage region is related to the terms in the curly brackets. Since the Debye length is proportional to the resistivity of the electrolytic solution, the relative change in voltage will decrease as the concentration of the lubricant additive increases. Figure 2 shows that this correlation holds true within the concentration ranges used for the experiments if the temperature and doping level are held constant. The correlation between these variables and the size of the damage zone has recently been reported [17].

The kinetics of the voltage change is related to adsorption and the variability with time of the conductivity of the damage region. The voltage varies with time as cracks propagate and expose new surface to the electrolyte. The mechanisms for the kinetic behavior are therefore tied to the increase of b and σ_2 with time. These mechanisms will be the subject of upcoming publications.

CONCLUSIONS

1. The in-situ electrical measurement technique provides information on the time dependent and steady state generation of mechanical damage in silicon during scratching by a dead-loaded spherical diamond. The relative changes in voltage (0.98 N) are 3.42, 6.17, and 3.75 % and the slopes are 106, 93 and 70 for the unlubricated surface, de-ionized water and 0.100 wt% lubricant additive respectively.

2. The change of voltage with time also depends on the dead-load on the diamond. The steady state voltage change increases as the load increases but the influence of lubricant is decreased.

3. The slope of the voltage versus time curve versus alkylated surfactant concentration is negative. This data is consistent with the model that the dielectric constant of the lubricant dominates the Debye length in the silicon.

4. The steady state variation in the voltage-time plot has been modeled as a damage region whose size and conductivity determines the voltage.

ACKNOWLEDGEMENTS

This work was supported partially by the National Science Foundation-Tribology Program under grant No. MSM-8714491. We thank Dr. Larsen-Basse for his support. Thanks are also extended to Dr. Jang Yul Park of Argonne National Laboratory for the conductivity measurements of the alkylated surfactants.

REFERENCES

1. A.R.C.Westwood, Phil. Mag., 17, p633, (1962)
2. S.Danyluk, LSA proc., JPL82-9, p89, (1981)
3. H.Foll, Appl. Phys. A53, p8, (1991)
4. S.Schaffer, Phys. Status Solidica, 19, p297, (1967)
5. M.Imai and K.Sumino, Phil. Mag., A47, p599, (1983)
6. D.J.H.Cockayne and A.Hons, J. de Physique, C6, p11, (1978)
7. S.Danyluk, S.W.Lee and J.Ahn, J. Appl. Phys., 63(9), p4568, (1988)
8. J.Ahn and S.Danyluk, Mat. Res. Soc. Symp. Proc., 140, p319, (1989)
9. R.Mouginot, J. of Mat. Sci., 22, p989, (1987)
10. A.Badrick and K.Puttick, J. Phys. D, 12, p909, (1979)
11. G.M.Hamilton, Proc. Instn. Mech. Engrs., 197C, p53, (1983)
12. W.T.Read, Phil. Mag., p12, (1955)

13.   E.Gerlach, J. Phys. C, 19, p4585, (1986)
14.   R.A.Vardanian and G.G.Kirakosian, Phys. Stat. Sol. B, 126, pK83, (1984)
15.   S.Jahanmir and M.Beltzer, ASLE Transactions, 29, p423, (1986)
16.   C.N.Rowe, Lubricated Wear, ASCE Trans., p143, (1987)
17.   S.Danyluk and S.W.Lee, J. Appl.Phys., 64(8), p4075, (1989)
18.   V.Celli and M.Kabler, Phys. Rev., 131, p58, (1963)
19.   F.Louchet and D.Cochet, Phil. Mag., A57, p327, (1988)
20.   P.J.Holmes, The Electrochemistry of Semiconductors, Academic Press, New York, (1967)
21.   M.Maeda, Structure and Properties of Metal Surfaces, Maruzen, Tokyo, (1973)

RECEIVED November 27, 1991

INDEXES

Author Index

Affiliation Index

Subject Index

Production: Donna Lucas
Indexing: Deborah H. Steiner
Acquisition: A. Maureen Rouhi and Anne Wilson
Cover design: Amy Hayes

Printed and bound by Maple Press, York, PA

Bestsellers from ACS Books

The ACS Style Guide: A Manual for Authors and Editors
Edited by Janet S. Dodd
264 pp; clothbound, ISBN 0–8412–0917–0; paperback, ISBN 0–8412–0943–X

Chemical Activities and Chemical Activities: Teacher Edition
By Christie L. Borgford and Lee R. Summerlin
330 pp; spiralbound, ISBN 0–8412–1417–4; teacher ed. ISBN 0–8412–1416–6

Chemical Demonstrations: A Sourcebook for Teachers,
Volumes 1 and 2, Second Edition
Volume 1 by Lee R. Summerlin and James L. Ealy, Jr.;
Vol. 1, 198 pp; spiralbound, ISBN 0–8412–1481–6;
Volume 2 by Lee R. Summerlin, Christie L. Borgford, and Julie B. Ealy
Vol. 2, 234 pp; spiralbound, ISBN 0–8412–1535–9

Writing the Laboratory Notebook
By Howard M. Kanare
145 pp; clothbound, ISBN 0–8412–0906–5; paperback, ISBN 0–8412–0933–2

Developing a Chemical Hygiene Plan
By Jay A. Young, Warren K. Kingsley, and George H. Wahl, Jr.
paperback, ISBN 0–8412–1876–5

Introduction to Microwave Sample Preparation: Theory and Practice
Edited by H. M. Kingston and Lois B. Jassie
263 pp; clothbound, ISBN 0–8412–1450–6

Principles of Environmental Sampling
Edited by Lawrence H. Keith
ACS Professional Reference Book; 458 pp;
clothbound; ISBN 0–8412–1173–6; paperback, ISBN 0–8412–1437–9

Biotechnology and Materials Science: Chemistry for the Future
Edited by Mary L. Good (Jacqueline K. Barton, Associate Editor)
135 pp; clothbound, ISBN 0–8412–1472–7; paperback, ISBN 0–8412–1473–5

Personal Computers for Scientists: A Byte at a Time
By Glenn I. Ouchi
276 pp; clothbound, ISBN 0–8412–1000–4; paperback, ISBN 0–8412–1001–2

Polymers in Aqueous Media: Performance Through Association
Edited by J. Edward Glass
Advances in Chemistry Series 223; 575 pp;
clothbound, ISBN 0–8412–1548–0

For further information and a free catalog of ACS books, contact:
American Chemical Society
Distribution Office, Department 225
1155 16th Street, NW, Washington, DC 20036
Telephone 800–227–5558

Highlights from ACS Books

Good Laboratory Practices: An Agrochemical Perspective
Edited by Willa Y. Garner and Maureen S. Barge
ACS Symposium Series No. 369; 168 pp; clothbound, ISBN 0–8412–1480–8

Silent Spring Revisited
Edited by Gino J. Marco, Robert M. Hollingworth, and William Durham
214 pp; clothbound, ISBN 0–8412–0980–4; paperback, ISBN 0–8412–0981–2

Insecticides of Plant Origin
Edited by J. T. Arnason, B. J. R. Philogène, and Peter Morand
ACS Symposium Series No. 387; 214 pp; clothbound, ISBN 0–8412–1569–3

Chemistry and Crime: From Sherlock Holmes to Today's Courtroom
Edited by Samuel M. Gerber
135 pp; clothbound, ISBN 0–8412–0784–4; paperback, ISBN 0–8412–0785–2

Handbook of Chemical Property Estimation Methods
By Warren J. Lyman, William F. Reehl, and David H. Rosenblatt
960 pp; clothbound, ISBN 0–8412–1761–0

The Beilstein Online Database: Implementation, Content, and Retrieval
Edited by Stephen R. Heller
ACS Symposium Series No. 436; 168 pp; clothbound, ISBN 0–8412–1862–5

Materials for Nonlinear Optics: Chemical Perspectives
Edited by Seth R. Marder, John E. Sohn, and Galen D. Stucky
ACS Symposium Series No. 455; 750 pp; clothbound; ISBN 0–8412–1939–7

Polymer Characterization:
Physical Property, Spectroscopic, and Chromatographic Methods
Edited by Clara D. Craver and Theodore Provder
Advances in Chemistry No. 227; 512 pp; clothbound, ISBN 0–8412–1651–7

From Caveman to Chemist: Circumstances and Achievements
By Hugh W. Salzberg
300 pp; clothbound, ISBN 0–8412–1786–6; paperback, ISBN 0–8412–1787–4

The Green Flame: Surviving Government Secrecy
By Andrew Dequasie
300 pp; clothbound, ISBN 0–8412–1857–9

For further information and a free catalog of ACS books, contact:
American Chemical Society
Distribution Office, Department 225
1155 16th Street, NW, Washington, DC 20036
Telephone 800–227–5558